Fundamentals of Case Management Practice

Exercises and Readings

Nancy Summers
Harrisburg Area Community College

BROOKS/COLE
™
THOMSON LEARNING

Australia • Canada • Mexico • Singapore • Spain
United Kingdom • United States

BROOKS/COLE
™
THOMSON LEARNING

Counseling Executive Editor: Lisa Gebo
Assistant Editor: Susan Wilson
Editorial Assistant: JoAnne von Zastrow
Marketing Manager: Caroline Concilla
Project Editor: Matt Stevens, Teri Hyde
Print Buyer: Tandra Jorgensen
Permissions Editor: Joohee Lee

Production Service: Gustafson Graphics
Copy Editor: Linda Ireland
Cover Designer: Bill Stanton
Cover Images: Courtesy of Eyewire
Compositor: Gustafson Graphics
Printer: Phoenix Color Corp.

Wadsworth/Thomson Learning
10 Davis Drive
Belmont, CA 94002-3098
USA

For more information about our products,
contact us:
Thomson Learning Academic Resource
Center
1-800-423-0563
http://www.wadsworth.com

International Headquarters
Thomson Learning
International Division
290 Harbor Drive, 2nd Floor
Stamford, CT 06902-7477
USA

UK/Europe/Middle East/South Africa
Thomson Learning
Berkshire House
168-173 High Holborn
London WC1V 7AA
United Kingdom

Asia
Thomson Learning
60 Albert Street, #15-01
Albert Complex
Singapore 189969

Canada
Nelson Thomson Learning
1120 Birchmount Road
Toronto, Ontario M1K 5G4
Canada

**Library of Congress Cataloging-in-
Publication Data**
Summers, Nancy.
    Fundamentals of case management practice :
exercises and readings / Nancy Summers.
        p. cm.
    Includes bibliographical references and
index.
    ISBN 0-534-35594-3
        1. Social case work—Problems, exercises,
etc. 2. Social work education. 3. Social
service—Problems, exercises, etc. I. Title.
HV11.5 .S86 2000
351.3'2—dc21   00-037969

To my parents, whose humor and wisdom about people and relationships
formed the foundation for my work with others

Table of Contents

Preface

In a small nonprofit agency handling cases of domestic violence, a woman answers the phone. She assesses the caller's concerns, accurately notes the caller's ambivalence on the inquiry record, and readily connects the caller to the person most able to assist. Down the street a young man acting as a case aid in a mental health center handles calls from clients assigned to the case managers. He competently notes the callers' mental status, and he asks the questions he knows will give him information doctors and therapists will need later as they work with these callers. His notes are clear and useful. How long did it take these two people to acquire these skills? Did they acquire this ability well after being hired in a social service agency, or did they arrive with a clear degree of competence?

PURPOSE

For me and for students, the issue has been how we can teach the skills that will promote their walking from the classroom into the social service setting with confidence. How can we be assured that students, often steeped in sound theoretical knowledge, will be able to fill out an inquiry form or make a referral effectively?

In addition, it seems important to equip students with the vocabulary and methods used by more advanced professionals in the human service field. While entry-level individuals would not give a *DSM-IV** diagnosis, it seems useful for individuals entering the field to be knowledgeable about what such a diagnosis is and what is meant by an Axis I or Axis II diagnosis. In this way, conversations among professionals would not be misunderstood.

Today individuals with a sparse education or with college degrees are finding themselves thrust immediately into roles for which there has been little formal training. It seems important, therefore, to find a method for teaching the actual human service experience at the entry level. *Fundamentals of Case Management Practice: Exercises and Readings* seeks to provide that experience in a thorough, step-by-step process that leads the reader from intake through monitoring to termination.

FORMAT

For each chapter in the workbook, basic information is laid out, followed by many exercises that prompt the reader to handle real issues and practice real skills. As readers progress through the text, they gradually assemble files on specific cases. Classroom discussion about these cases and the best disposition for each of them are not unlike those discussions that occur every day in a variety of social service settings.

The textbook also contains a set of forms that can be copied. These forms, taken from actual social service settings, give the reader an opportunity to practice accuracy and skill in handling social service forms and records.

ORGANIZATION OF THE TEXTBOOK

In Section 1, "Foundations for Best Practice in Case Management," readers are introduced to important foundation pieces. Ethics and ethical issues, the reasons for case management, and the importance of the ecological model in assessment and planning give readers an introduction to professional basics.

---

*Note that *DSM-IV* was revised in 2001 to become *DSM-IV-TR.*

In Section 2, "Useful Clarification and Attitudes," readers are invited to examine what in their thinking will impede effective helping in the social service setting. Beginning with issues of cultural diversity and moving to the role of personal attitudes, this section concludes with information and exercises related to determining who owns the problem. Each chapter in this section contains exercises encouraging readers to examine realistically their own attitudes and judgments.

Section 3, "Effective Communication," begins by introducing the reader to good and poor responses with exercises that help readers see the consequences of poor communication. Chapters on asking questions, responding to emotions, confronting problematic behavior, and disarming anger are included. The section ends with exercises designed to have students practice all the communication skills in order to smooth out the communication and allow it to become natural and responsive.

In Section 4, "Meeting Clients and Assessing Their Strengths and Needs," readers begin to take inquiries for services. Forms are provided that ask for basic information, teaching the reader what is important to find out in that first call. This section also includes a chapter on preparing for the first interview, helping the reader become sensitive to issues the client might have at a first meeting.

Introductions to the *DSM-IV* and to the mental status examination allow the reader to become familiar with the vocabulary and the information most important to other professionals in the human service field. The reader is encouraged to begin noting how a person seems to them at the time of contact. The chapter and classroom discussion help students pin down what is important to note.

In this section readers also practice completing release of information forms for the clients they have developed in the classroom setting, mastering which records are useful and which are not.

Section 5, "Developing a Plan with the Client," allows readers to further develop those clients for whom they have created phone inquiries. Here, individually or in planning teams, according to the instructor's process, students develop realistic plans for their clients. A chapter is included instructing readers on how to prepare for and participate in team planning. In the final chapter, students refer cases to providers of service.

The final section, Section 6, "Monitoring Services and Following the Client," begins with a switch to the role of a provider of service where students take general goals given them by case managers and develop specific goals and objectives within stipulated time lines.

In this section readers also learn the importance of monitoring cases from a case management perspective, how to document contacts related to the client, and how to terminate the case. There are numerous documentation exercises, an opportunity for students to begin to write professional notes and keep good records.

TO THE STUDENTS

It is always a challenge to give students the skills and information they will need on the first day of their first job. Even when students are already working in the field and managing many of the tasks well, they often do not know for certain why agencies choose to do things one way as opposed to another. This textbook seeks to empower you to be able to function competently and to know why you are proceeding or should be proceeding with clients in a particular way.

In *Fundamentals of Case Management*, Practice you will follow a specific series of steps, beginning with what you are thinking and how you think ethically in client-worker relationships, continuing through your communication with your clients, and ending with your putting together case files and managing cases.

Throughout the course you will find yourself in discussions with others about possible treatment or service plans or the dynamics of a person's situation. Use these to learn more about collaboration and to increase your ability to participate in the same sort of discussions in the agency where you will go to work.

Many students have taken this textbook to work with them and have found it both useful and realistic. Students have contributed their experiences on the job to make this textbook replicate as nearly as possible the issues and concerns you will encounter in your work with other people.

TO THE INSTRUCTOR: SUGGESTIONS FOR USING THIS TEXT

This text can be used to take students step-by-step through the case management process outside of the often harried and pressured atmosphere of a real social service agency. When the student is ultimately confronted with the actual situation, the routine and expectations will not be new. Chapters are broken down into each step in the process. Students progress in their skill level, finally creating cases and caseloads with you acting as the supervisor, much as a supervisor would act in an actual agency. Without the urgency, you will have time to let students look up information, discuss possible diagnoses, and develop sound interventions under your guidance. For example, exercises on DSM-IV-TR or the mental status exam may have a number of possible answers. Your discussion with your students, similar to the discussions that take place in agencies about these possibilities, is more important than the actual answer that is chosen.

Every chapter includes exercises to help students practice their skills. Often there are several versions of the same exercise. It is useful to students to begin in small groups to address the issues posed in the exercises. Their discussions and the ideas and concerns they bring back to the larger class are consistent with discussions held in social service agencies. Later versions of the exercises can be used as tests, or you can go back to them at a later time to make sure students continue to practice their skills.

BENEFITS AND ADVANTAGES

This material has been used in my own classroom for ten years, updated to meet current social service trends. Students have commented that using this text is like walking from the classroom into the social service setting with very little lost time in learning the actual process. Instructors teaching the practicum course have used the word *empowered* when describing what this text has done to give students confidence and skill in their first encounter with a social service position.

There are three positive features of this textbook that make it especially useful in preparing students to work in this field.

- First, the text gives very basic information a person would need to handle each of the tasks described. Theoretical information can be found in many other places, and thus the concentration has been to focus on what is important to note, think about, document, or pass on in each step of the human service process.
- In addition, there are numerous exercises creating very real situations for students to consider and handle. These exercises are based on real experiences taken from my 23 years of practice in human services and from the experiences of many others who graciously contributed to this book. Doing the exercises and participating in the classroom discussions that follow will expose the student to an extremely broad range of possible circumstances and difficulties in the field.
- Finally, the book contains forms that give students an opportunity to practice compiling information at various times throughout the management of the case. These forms can be copied and used to create files on clients developed by the student. Using each form a number of times gives students practice in preparation for real clients in real social service settings.

These features, when taken together, create a nearly realistic social service setting in the classroom, giving the instructor many opportunities to strengthen student skills and sensitivity. In addition, a forthcoming textbook, *Fundamentals for Practice with High Risk Populations*, will support students with applicable details and considerable information on various at-risk populations. When this textbook is available, students can use the material found in this supplement to develop realistic clients, create useful service plans, and make appropriate referrals.

ACKNOWLEDGMENTS

I would like to thank Dahl Krow, executive director of the Dauphin County Mental Health and Mental Retardation Case Management Unit, and the entire staff who graciously spent time with me describing their concerns and process and providing me with forms and anecdotes. In particular I want to thank Deborah Gantt, a mental health supervisor, supervising targeted case management, and Kelly Preston, a mental health supervisor, supervising the intake function of the agency. Both of them provided valuable information that made this book more relevant and useful.

I would also like to thank the reviewers of this workbook for their helpful comments and suggestions: Mark Homan, Pima County Community College; Bruce Friedman, Wayne Community College; Diane Coursol, Mankato State University; Tricia McClam, University of Tennessee, Knoxville; Alisabeth Buck, Tacoma Community College; Mary Davidson, Columbia-Greene Community College; Mary Kay Kreider, St. Louis Community College-Meamec; Mary DiGiovanni, North Essex Community College; Thomas Collins, University of Scranton; and Joseph Walsh, Virginia Commonwealth University.

Foundations for Best Practice in Case Management

Chapter 1

Ethics and Other Professional Responsibilities for Human Service Workers

INTRODUCTION

Ethical principles are the foundation of good human service practice. Each social welfare profession, from psychologists to social workers to human service workers, develops a set of ethical principles appropriate to the practice. Most professions monitor the behavior of their members with regard to these principles, singling out those who violate ethics for disciplinary measures.

Ethical principles are generally created in order to protect clients of the profession from exploitation. In the work we do, there is considerable opportunity to exploit vulnerable people, since the clients who seek our help are dependent upon us for the aid they need. Any violation of their trust on our part will only compound the clients' problems. But . . . when professionals practice within the parameters of ethical principles, the public can feel confident that their interests will be respected and protected. Thus, ethical principles inform the decisions we make that affect clients, and they provide guidance in choosing the approaches we take with clients.

In this chapter we will look at some ethical guidelines common to all of the helping professions. Failure to know and follow these guidelines in your future practice of human services can result in your dismissal from an agency or, worse yet, in a civil suit brought against you for a violation of the ethical code wherein the violation caused damage to the client. Although violations of ethical principles may cause extremely negative consequences to you and your career, they are always extremely destructive for the client, who is already vulnerable.

DUAL RELATIONSHIPS

A dual relationship occurs when you and a client to whom you are giving services have more than one relationship. You may be the client's case manager and her cousin, her boyfriend, or her customer at her beauty salon. You may be his case manager and his employer for your yard work, his Sunday school teacher, or his Little League coach.

The first rule is to avoid all dual relationships. Your practice gives you a position of power. The clients look up to you as someone who can provide real assistance.

Furthermore, you might be the one who will determine when a client can return to work, or you may be the person who reports a client's attendance in your program—attendance that keeps that client out of jail. It is possible that you could exploit or give the appearance of exploiting this power. In addition, there is enormous potential for a conflict of interest.

Suppose, for example, that your supervisor tells you on Thursday afternoon that you have been chosen to represent the agency at a big dinner being given to honor a county official at the Hilton Hotel on Saturday night. This gives you little time to prepare. You need to get your hair cut and styled. You call your client, who is a hair stylist, and prevail on him to work you in at the last moment. Your client does you a favor and sees that you get a good appointment. You are very grateful, go to the wonderful dinner, and think little more about it.

Several months later your client calls you. He has a need for a prescription refill from his psychiatrist. It is Friday afternoon, and the psychiatrist will not be back in the agency until the following Wednesday. The client feels you should be able to do this favor for him because of the favor he performed for you. He does not have time to see the psychiatrist regularly, he tells you. He cannot understand why you are being "so rigid" about calling in a prescription for him without the doctor's prior knowledge. He indicates that he thought the two of you were friends who helped each other out when needed.

Whatever you do in this situation, you will lose. If you call in the prescription, you have, at worst, violated an agency rule that a client must be seen at regular intervals by his psychiatrist before medications can be refilled. This could cost you your position or result in a disciplinary action. On top of that, you could start down a very slippery slope with this client in which he comes to expect special favors of you and offers you, in return, special, very tempting favors related to his business. On the other hand, if you do not call in the prescription, you have alienated a client who needs the services of your agency. You have created a barrier to his feeling comfortable with you and getting the help he needs in the future. The client is harmed. A relationship that was or could have been useful to him in resolving problems is now something else. The opportunity for real progress is diluted with issues of friendship and favoritism. From the short-sighted point of view, each of you may see the convenience of exchanging favors as trivial and unrelated to the therapeutic relationship. In the long run, however, the relationship can never return to a professional one; and if, in the future, your client is in acute need, you may no longer be able to provide the professional intervention he needs.

In some very small, rural communities, it is not possible to avoid dual relationships entirely. In those situations, after doing all that you can to make other arrangements, you must talk with the client about the possible problems that could arise and how each of you must avoid these problems together. The client then has the choice to continue the relationship, find other arrangements, or discontinue services altogether.

Although the situation of gifts from clients does not pose a dual relationship, clients who bring gifts for you do pose a particular conflict of interest if you accept the gifts. Keep the relationship professional, and try to avoid accepting gifts. Often, though not always, gifts are the client's way of manipulating the situation. "I'll give you this item and you accept it. Next time you owe me something" or "I gave you this lovely thing. I am such a nice person. Even you think so or you would not have taken my gift. How can you then refuse to give me what I want?" Clients need to learn to express their desires clearly rather than by using gifts.

Gifts, however, are not always manipulative. A case manager working in a fuel assistance program worked closely with a family. The husband was injured when an automobile he was working on at a garage slipped on the lift. Unable to work, the family's meager resources began to dry up. The wife managed to find work in a greenhouse but, as winter approached, was laid off; and the expenses, particularly for fuel, increased.

The couple had two children, both in elementary school, and the couple struggled to clothe and feed them as the wife sought another job.

The case manager saw them through this difficulty, getting the family welfare checks, seeing that the husband enrolled in the community college for courses in high-tech auto repair while his injuries healed, and finding school clothes for the children. The husband did well that winter in school and set a good example for his children, who seemed to do better in school than they had the previous year. The wife returned to the greenhouse in the spring and found that she was needed as a manager, with the strong possibility that she would have a year-round position there.

Elated by how well things were going and how much better the future looked, the couple came to see the case manager one day in the early summer and brought her a pot of black-eyed susans from the greenhouse. "We just wanted you to have these for all you have done for our family," the husband said, smiling expansively. The husband and wife looked pleased and happy. Obviously the couple was proud to now be in the position to able to give something too. It was important to them not to see themselves as the recipients of handouts all the time, but to be able to also give something to someone who had been helpful.

Refusing a gift in such circumstances can be interpreted as rejection. If the worker had said, "Oh, I can't accept that. You'll have to give it to someone else," the client might have heard a different message: "You are the client and I am the benevolent worker. I help you, but you can never get to the position where you could possibly do anything for me. I don't need anything you could give me; but you, on the other hand, are a poor soul in need of my help."

If your agency has a policy against your personally accepting gifts, try to find a way to accept a gift of this sort on behalf of the agency. In this case, the worker planted the flowers in a planter near the door of the agency. This is a better solution than outright rejection of the gift.

The rule is to be very careful about accepting a gift from a client. Note in the record the client's offer, and whether the gift was accepted or rejected and why.

Sexual or Romantic Relationships

Individuals who come to us as clients often feel isolated, discouraged, and misunderstood. The relationship they form with a respectful, concerned human service worker may be the first time they have felt appreciated. It may be that this relationship is so comforting to them that they attempt to turn it into something more permanent, more personally meaningful.

It is not uncommon for some clients to fall in love with their workers in what we call transference. In a sense, these clients transfer to their workers the attributes they are seeking in another person. They may assume the love and affection they are seeking will be forthcoming from their worker because the worker appears to be so kind and helpful. The case manager's encouragement may be the first time a client has felt appreciated or understood. This happens because of an erroneous perception on the part of the client. Concern and encouragement are seen as gestures of love and affection, an invitation to create more than a professional relationship.

There is also countertransference. It is not unusual for a worker who is harried and overworked, possibly coping with difficulties in his or her personal life, to find the willing ear of a client very supportive. Clients are often attractive, sensitive people and can convey warmth and support when a case manager is most vulnerable.

Take, for example, Kent, a case manager. Kent's wife left, in the middle of a Tuesday morning, and Kent was to be at work that afternoon at 3:00. After endless fighting and his trying to dissuade her from going, she left. For the rest of the day before work, Kent tried to get some money from their joint account, tried to find out where his wife was going, and tried to make some decisions. He arrived at work feeling exhausted and bitterly betrayed.

That evening, Kent made a home visit at Lucy's house. Lucy had first come to the agency with extreme depression, but was doing so well now that it was thought that these follow-up visits might be terminated. Lucy, an artist, greeted Kent warmly. She had put on a pot of tea and made some banana bread for his visit. Gratefully Kent sank down on her sofa. Instead of asking Lucy how things had been going and whether she had enough medication or if she had any medication questions, Kent found himself talking about his upsetting day.

In response to Lucy's first remark, "You don't look very well tonight, Mr. Paulman," Kent heard himself pour out the day's events; and then he went on to talk at length about how his marriage had unraveled. He felt comforted by Lucy's interest in him as she listened intently. Here was a person who appeared to respect him as a professional and as a person. Here was a woman willing to listen to his problems. Here was a warm retreat from the job and the problems of the day where Kent could feel safe and supported.

Kent never intended for a real relationship to develop between Lucy and himself. In fact, as he left that night, he told himself that he might have crossed a dangerous line and that he should avoid further contact of this sort with Lucy. Nevertheless, based on that evening, Lucy called; and because Kent was lonely, his life was uncertain, and he was filled with anger and bitterness about his situation, he continued to find in Lucy a warm, supportive person, someone who could reassure him by her presence that he was attractive and interesting.

The relationship moved from his visiting in her home after work to his staying overnight at her house to his moving in his belongings and beginning to live there. They went out on dates. The furtiveness of these activities only made the relationship seem more romantic and important. Finally, a supervisor discovered the relationship, and Kent lost his job. After three years, he and Lucy have separated, and Kent is not working in the human service field anymore because he violated such an essential ethic. Instead, he sells appliances in a local store. Lucy became depressed when the relationship ended and has entered treatment again.

The person responsible for maintaining a professional relationship regardless of personal feelings is the case manager. Regardless of how you feel about your client or how the client apparently feels about you, you are the responsible party. You will be penalized if the relationship crosses from professional to intimate. It is always assumed that the client is the vulnerable party.

Here are some warning signs based on a list created by attorney O. Brandt Caudill (1996).*

Warning Signs from the Client

- The client shows overt sexual interest in the worker either through conduct or verbally.
- The client describes dreams that are increasingly sexual in which the worker is prominently involved.
- The client is excessively interested in the worker's private life.

Warning Signs from the Worker

- The client is prominent in the worker's dreams.
- The worker looks forward to seeing the client, more so than other clients.
- The worker begins to see the client as more understanding than others in the worker's life.

*Used with permission.

- The client inquires about the worker's relationship with his or her spouse and children.
- The client attempts to give the worker romantic gifts. Be careful about accepting gifts from a client, and note in the record whether the gift was accepted or rejected and why.
- The client wants to see the worker outside the office in places such as restaurants or movie theaters.
- The client gives the worker romantic poetry or brings in romantic articles and books.
- The client dresses in seductive attire.
- The client interprets the worker's statements of concern for the client to mean the worker has a romantic interest in the client.
- The client repeatedly hugs and touches the worker.
- The client indicates a desire to be special to the worker.

- The worker inquires about the client's sexual life and fantasies when these are not relevant to case management.
- The worker is more interested in this client's attire than in the attire of other clients.
- The worker is more concerned with his or her own attire on days when he or she will see the client.
- The worker begins to see the client as a person without issues or problems or minimizes these so that the client seems more acceptable as a partner or friend.
- The worker takes many innocuous actions the client might make to mean the client has a special interest in him or her.

Be Sure to Note!

It is a violation of all ethical codes, and in most states against the law, to engage in a sexual or romantic relationship with a client. This is *clearly* exploitation. It is never tolerated. You must be aware that attractions can occur between clients and those who provide service, but that acting on those attractions is entirely unethical in professional practice, and illegal in most states as well.

VALUE CONFLICTS

Generally, a client's values and your own have little to do with why the client is seeking services from you. Sometimes, however, religious, moral, and political values play a pivotal role in the problem the client is bringing to your agency. It would be rare for a case manager to get deeply involved in these primary problems, but it could conceivably happen.

First, you begin by consciously knowing yourself and your feelings about certain value-laden issues. If a severe conflict of values occurs, you should be able to tell the client that the conflict exists and that it could interfere with services. In order to begin this process, complete the self-assessment exercise in Figure 1.1. Here you can begin to inventory some of your own attitudes and strong feelings.

Second, if the conflict is severe, you might need to make arrangements to transfer the client to another case manager. You would not do this if it were a simple value conflict, but you would try to make the transfer if you found you could no longer be objective, you were extremely uncomfortable with the client because of her values, or you felt compelled to counteract the client's values by imposing your own.

For example, a case manager did not personally believe in birth control, including tubal ligation. She was a case manager for individuals with developmental disabilities. When a young couple on her caseload decided to marry, she became actively involved in discouraging this, particularly when she learned that the woman planned to have a tubal ligation so that they would not have children. The families of the two individuals supported the marriage. The clients were high functioning; each client had a job, and each had the support of other community agencies.

In the months before the wedding, the case manager did not attempt to transfer the cases to another case manager. Instead she harangued the couple about the sins of birth control and of marriage without children, and about the unwise decision to marry at all, given their "mental impairment." The families complained to the agency, asking that she stop pressuring these vulnerable individuals. Twice the supervisor disciplined the worker. When the worker persisted—visiting the couple's minister who would perform the ceremony, the supervisor where the man worked, and the woman's parents—she was fired from her position.

For the clients, two people who had always relied on those who seemed to be wise, the worker introduced uncertainty and fear. Her constant negative warnings damaged their fragile self-confidence and self-esteem. For them, what should have been a happy decision, supported by family and friends, became a decision fraught with anxiety. Family and friends had to work long and hard to restore their confidence in their original decision.

This is an example of the worst possible way to handle a values conflict. In this situation, the worker attempted to impose her own point of view, her own personal values, on the clients. She denied them the right to choose for themselves and interfered in what was largely a personal and family issue unrelated to case management.

Avoiding Value Conflicts

Here are some rules for avoiding value conflicts and ensuring the client gets professional service:

1. Be respectful of attitudes and lifestyles that differ from your own.
2. Never practice prejudice toward minorities, the handicapped, or those differing in sexual preference.
3. Never refrain from giving your best service to a person even when you disagree with the person.
4. Never attempt to change the consumer's values to coincide with your own.

CLIENTS' RIGHTS

Anyone who gives service in one of the helping professions must be familiar with the rights of the client and make a particular effort to see that clients understand they have rights when they seek help. Often clients mistakenly assume that they have few or no rights when they come in for services. In addition, professionals may fail to inform clients of their rights because it is easier to work with individuals who are vulnerable, dependent, and uninformed. This, of course, sets up a situation in which it is easy to exploit the client. The purpose of educating clients about their rights is to allow them the opportunity to become active participants in their care and partners in decisions that affect them.

A Self-assessment exercise—Possible Values Conflicts When Helping Others

Look at each description of a person or group of people, and assign a number to each.

1 Give the description a 1 if you think you could work with the person or group.

2 Give the description a 2 if you think you could work with the person or group, but would find it uncomfortable or difficult.

3 Give the description a 3 if you could not work with the person at all.

_1_ 1. A woman who wants you to help her feel comfortable with her decision to have an abortion

_1_ 2. A man who frequently brings up his fundamentalist religious beliefs

_1_ 3. A homosexual couple who wants help in improving their relationship and resolving their interpersonal conflicts

_1_ 4. An interracial couple seeking premarital counseling

_2_ 5. A man from Iran who strongly opposes the equality of women and talks about women working in denigrating tones

_2_ 6. A man who has for years been getting more welfare than he is entitled to receive by using certain tricks he developed to beat the system

_1_ 7. A man and woman who say they want to improve their marriage, but the man will not end his affair with a second woman

_1_ 8. A white couple seeking help for behavior problems with their adopted son, who is African American

_1_ 9. A man who makes it clear he disciplines his children by using corporal punishment often

_2_ 10. A person who refuses to discuss feelings and says that all that matters is facts and logic

_1_ 11. A woman who has chosen prostitution as a way to support herself and her children

_1_ 12. A gay man dying of AIDS who comes in with his lover to resolve conflicts around how he contracted the disease

_1_ 13. A man seeking help to curb his extreme abuse of his wife

_1_ 14. A woman who sexually molested her son

_1_ 15. A lesbian couple seeking to adopt a child

_2_ 16. A woman dying of breast cancer who wants to take her own life

_1_ 17. A person who relies heavily on cocaine to get through the day

_1_ 18. A couple who is openly anti-Semitic

_2_ 19. A vocal member of the Ku Klux Klan

_1_ 20. An orthodox Muslim who cannot always see you because his appointments often interfere with his times of prayer

Figure 1-1

Most agencies prepare client rights handbooks for clients to keep as a reference. A hospital for the mentally ill would include the right to be released from the hospital as soon as care and treatment in that setting are no longer required. Some agencies inform clients that they have the right to participate in the development and review of their treatment plan. Clients generally have the right to participate in major decisions affecting their care and treatment. Most clients who are involuntarily committed to an inpatient setting have the right to refuse treatment to the extent permitted by laws in that state or the right not to be transferred to another facility without clear explanations regarding the need for the transfer. Inpatient units stipulate it is the client's right not to be subjected to harsh or unusual treatment. The hospital may also spell out the fact that the client may keep and use personal possessions, or the client must be informed about why something is being removed. In most settings, the client will have the right to handle personal affairs and to practice the religion of his or her choice.

In outpatient settings, a client usually will have the right to flexible and responsive treatment plans, the right to expect an individualized plan of service, and the right to make suggestions and express concerns. Often there is a procedure clients can follow if they are dissatisfied with the worker assigned to them.

The following sections discuss some important rights that belong to clients.

Informed Consent

A client receiving services always has the right to consent to these services or withdraw from them. In making this decision, the client must be informed enough to make a wise decision. When the client is informed and consents to treatment, we call that informed consent. Making certain that a client can give informed consent begins with the intake, where agency policies are explained and choices of treatment or services are outlined. This level of information should continue throughout the entire relationship between the client and the agency until termination. This means that clients informed about treatment or services can make their own decisions with regard to the services. The following list contains items that should be addressed when relevant to the client's services. The client has the right to be informed about:

1. Any side effects, adverse effects, or negative consequences that could occur as a result of treatment, medications, or procedures
2. Any risks that might occur if the client elects not to follow through with treatment or services
3. What is being offered to the client, including what the treatment is, what will be included, and any potential risks and benefits
4. Any alternate procedures that are available

Some of the people with whom we work have a limited capacity to understand all the details of service and treatment. It is your task to find an appropriate balance between too much and too little information and to make your information clear and easy to understand.

There are three parts or criteria that constitute informed consent. All must be present in order to say that the client gave informed consent. They are:

1. *Capacity.* The client has the ability or capacity to make clear, competent decisions in his or her own behalf.
2. *Comprehension of information.* The client clearly understands what is being told to him or her. In order to make sure that this is so, give your information carefully and always check to be sure the client understands what you have told him or her.

3. *Voluntariness*. The client gives his or her consent freely with no coercion or pressure from the agency or the professional offering the service.

Currently laws and courts are recognizing more and more often the client's right to self-determination. When we fail to tell our clients the information they need in order to give informed consent, we run the risk of being found negligent, particularly if the treatment or service involved was unusual.

Confidentiality

This is both an ethical principle and a legal one. It is the most basic right of any client, either in treatment or receiving services, to know that what the client is sharing in your office will remain confidential. It is important to protect the client from disclosures about his or her personal situation without the client having authorized such a disclosure.

Release of Information Form

Not many years ago a person seeking services, particularly from a public agency, signed a blanket permission statement allowing information to be shared with others as the agency and the worker saw fit. Today, release of information forms must state specifically to whom the information is being released and must be time-limited (good for three months, one year, and so on). Do not use forms that are not specific in this manner.

In addition, in most states, release of information forms must specifically state that you may release information regarding the client's HIV status. All references to HIV/AIDS must be deleted from the record unless the client signs a form that specifically states that you have permission to release this information. If you are asked to release information about a person who is HIV+ and the client signs a release form, the law in most states specifies that it is not good enough to simply remind the client that his case contains references to his HIV status and get his verbal permission to release the information anyway. You also must have his written permission. If the client has not given you written permission, you must delete all references to HIV/AIDS, including the fact that he may have been tested and the test was negative.

If your state does not have such a law, you are still responsible for protecting your client and must be alert to the possible harm such a release might have on the client. In such a situation, it is wise to involve the client in a discussion about the release of this sort of information or, if the client is unable to participate in such a discussion, to take steps to protect the client from undue bias.

In some instances, workers have informally notified their friends and acquaintances in other agencies of a client's HIV+ status, thinking they were doing these people a favor. In fact, this behavior is entirely unethical and can lull other workers into believing they know who is and who is not HIV+. We can never actually know this for certain because of the length of time it takes for the disease to register positive on a blood test. A person can be positive early in the illness and still have negative blood tests. For this reason, workers should use universal precautions with every client when those precautions are called for. A worker who fails to use universal precautions on the false assumption that he knows the client is not HIV+ places himself at undue risk.

Assignment

Find out what the laws are in your state for releasing information about a client that contains references to the client's HIV status.

Collegial Sharing

In addition, you should, out of respect for the client, ask the client's permission before sharing information with a colleague from whom you are getting an opinion or supervision, but not if the case will come up in the normal course of your supervisory meetings with your regular supervisor. Likewise, you cannot share information with a student intern without making certain the student has signed an agreement to observe strict confidentiality while part of the agency. Suppose you are working in an agency and have been asked to give a student a view of what you do. In order to illustrate what you have told the student, you show her several case files. She reads the cases and discovers that one of them is the boyfriend of her cousin. What she reads in the file is alarming to her, and she decides her cousin should not be dating the client. She leaves the agency and begins to share information with the cousin, causing considerable conflict among family members and anguish to the cousin who knew part of the story, but not all of it. This sharing of information is unacceptable, and most agencies do not allow students or volunteers to read anything before they have signed a pledge to honor the confidentiality of the clients.

Guarding Confidentiality on the Phone

There are other opportunities to violate confidentiality. For instance, a person receiving services from your agency may also be receiving services from a local physician. Suppose someone calls, claiming to be the physician's nurse and needing to know at once what medications the client is taking. She may really be the physician's nurse, or she may be a person posing as the nurse in order to determine that the client is using your services and the level of his or her problem. Even if she is the nurse, the client may wish to keep his or her physician uninformed about the involvement with your agency. All agencies have procedures for such situations in the event of a real emergency. You, however, must never openly and automatically acknowledge that a client is being seen in your agency, no matter how important and official the other person seems to be. In the case of a seeming emergency, refer the call to your supervisor.

Here is a possible scenario that is not an emergency:

    You: Hello.
    Caller: Hi. This is Ann Taylor. I'm a counselor at Harrisburg Middle
             School, and I'm calling about Jimmy Smith. Did he and his mother
             keep their appointment with you today?
    You: I'm sorry I can't help you with that. If you would have Mrs. Smith
             sign a release of information form stating what it is you need to
             know, we can send you that information, if she is known to us.

Another way to violate confidentiality is to talk about your cases with your friends and relatives, leaving out the names. Others may be able to piece together the identity of the person you are talking about based on other information they possess. In this way, they may discover far more about the client than the client ever intended them to know.

Minors and the Infirm

Take special care to protect the confidentiality of minors and the infirm (individuals who are frail, sick, and unable to fully participate in decisions). Not all systems respect confidentiality to the degree that we are committed to it.

In one children's case management unit, parents were routinely urged to sign blanket release of information forms. When the school requested information on a child being seen, all the information was sent to the school. It was stamped in red letters with the word *confidential*, and it was sent to the school psychologist. Nevertheless, school clerical personnel assisted in typing and filing information for the psychologist and generally read the information sent by the case management unit. Having no training in confidentiality, these clerical people talked among themselves about students, sharing personal information they had learned. Many times they passed on to teachers tidbits of what amounted to gossip. This information shared outside the professional context and without professional understanding jeopardized the progress of the children and the relationship of their parents with the school personnel. As these children moved through the school system, the gossip followed them. Keep an eye on what information is released, remembering that information given about a child can follow that child all through school, prejudicing the response to that child.

In another case, a woman with a developmental disability got a job at the police department as a cleaning woman. She was told that she needed to bring in her "records from mental health" so the police could know why she went there. She arrived at the case management unit, pleased about the job and ready to give all her records away. The case manager talked to her about the wisdom of retaining most of the information as confidential. In the end, a short statement was released, with the client's permission, giving only the most general information about her relationship with the mental health/mental retardation case management unit. It is important to remember that older people or individuals who do not have the capacity to protect themselves can be easily led to sign releases regarding information that might best be kept confidential.

When You Can Break Confidentiality

The law in all states does make exceptions. The following are circumstances that allow you to break confidentiality:

1. When you must warn and protect others from possible harmful actions by the client. For instance, you or your agency must warn another party if your client is intent on harming that other party. In addition, you should notify the police.
2. When the client needs professional services. For instance, if the client has taken an overdose of medication and is in the emergency room (ER), the ER staff may call, needing to know what prescriptions the client was taking in order to give the proper antidote.
3. When you must protect clients from harming themselves. An example might be people who are threatening to take an overdose of their medications with the intention of committing suicide or people who appear so depressed or desperate that they are talking about ending their lives.
4. When you are attempting to obtain payment for services and the payment has not been made. Your agency would refer a client for nonpayment only after reasonable attempts had been made to remind the client of this obligation and only if the client had made no effort to arrange even minimal payment.
5. When obtaining a professional consultation regarding how best to proceed with a case.

Privacy

Privacy is very much related to confidentiality. Siegel (1979) calls it "the freedom of individuals to choose for themselves the time and the circumstances under which and the extent to which their beliefs, behaviors, and opinions are to be shared." Stadler (1990) calls it "the right of persons to choose what others may know about them and under what circumstances."

Privacy is invaded or altered under some circumstances, and clients need to be informed of those circumstances. The point you should stress with your clients is the fact that third-party payors will have access to diagnoses and, in some cases, to actual records or summaries of records. The agency must provide this access in order to be paid for the services it has rendered. Many clients are unaware of this fact or unaware of the extent of the information being shared. They should have this situation explained to them.

Accessing the File

Most codes of ethics do not specifically discuss clients' right to read their files. This area can present a conflict for the worker who feels some reluctance about sharing all the notes and a diagnosis with the client, who may distort or be unable to understand fully what has been written. Opinion is divided on the issue. Some believe the clients are legally entitled to read their charts, and others believe it is better to explain the diagnosis and the notes in the chart to clients instead. Currently there is a strong consumers' rights movement that supports clients' right to see their files. In addition, there is evidence that clients who have read their charts and received clear information are less likely to sue for malpractice or create other legal problems.

If you allow a person to read the chart, be sure to sit and answer questions and explain what has been written. It is not a good idea to just hand someone a chart and provide no explanations for technical information that may be written there. This potentially creates misunderstanding. Sit with the client and carefully review the important points in the chart.

Self-Determination

Educating clients and informing clients about their rights are both done so that clients can exercise the right to self-determination. Paramount to any relationship between professionals and clients is the right to self-determination. Clients have the right to do research about their diagnosis or problem and to question the treatment plan or make suggestions. Clients have the right to withdraw from treatments and services they find are not helpful. Clients have the right to decide when and for how long they will use services (unless their involvement with the agency is based on an involuntary court commitment). Clients have the right to choose their own goals.

Often this presents a problem for a worker who feels compelled to look after the best interests of the client. One of the hardest lessons you will ever learn is how to let clients make mistakes and learn from those mistakes. You can make suggestions and express concerns, but ultimately clients have the right to determine what they will do.

You may feel strongly, for example, that one of your clients is not ready to walk away from the agency; and you may feel certain that the client's doing so prematurely will result in further problems with alcohol. In fact, when your client does leave treatment against your advice, she does eventually end up with another DUI charge. Although all your worst fears and predictions came true, you cannot know for sure that the work with you and the new charge were not important learning opportunities. In other words, clients have the right to test the waters, so to speak, and to learn that they are not as ready as they thought they were.

PRIVILEGED COMMUNICATION

Clients and workers alike talk about privileged communication without truly knowing what it is. First of all, it is a *legal concept*. It protects the right of a client to withhold information in a court proceeding. It is a right that belongs to the client. It does not belong to the worker or the agency.

All states have a law that stipulates what communication between a client and professional shall be considered privileged in order to protect the client from the disclosure of confidential information during a court proceeding. These laws designate who is to be considered a professional. You may recall a recent case in which a man who committed a murder confessed this murder in an Alcoholics Anonymous (AA) group. He tried to invoke the right of privileged communication, but the courts denied it because the state law did not specifically name AA as a group protected by this statute.

Only the client can invoke privileged communication in order to protect himself or herself. Professionals and agencies cannot use it to protect themselves. If the client waives the right to privileged communication, the professional or agency has no ground to withhold information. Clients waive this right if they sue your agency or if they use their condition as a defense in a legal proceeding.

WHEN YOU CAN GIVE INFORMATION

There are times when you may give information. They are:

1. If you are acting in a court-appointed capacity, such as that of guardian or payee
2. When you believe the client intends to commit suicide
3. If you or your agency is sued for malpractice
4. When a child under 16 years old is believed to be the victim of a crime such as sexual or physical abuse or sexual exploitation
5. When you determine the client needs to be hospitalized for a mental condition
6. When the court mandates that you turn over certain information
7. When the client has told you of his intention to commit a crime, harm another person, or harm himself
8. When the client uses a mental condition as her defense or as a claim in a civil action

The Intention to Harm Another

On October 27, 1969, Prosenjit Poddar killed Tatiana Tarasoff and set in motion court proceedings that brought about changes in the way confidentiality is viewed. Poddar was a patient of Dr. Lawrence Moore at Cowell Memorial Hospital at the University of California in Berkeley in October of 1969. Moore, a psychologist, was told by Poddar of his intention to kill Tatiana Tarasoff. Moore contacted campus police, who briefly detained Poddar but released him when he appeared to the police to be rational. Apparently, Dr. Powelson, Moore's supervisor, directed that no further action be taken to detain Poddar. No one warned Tatiana Tarasoff of the danger she faced.

The Tarasoffs brought charges against the professionals in this case for failure to warn the victim of the impending danger. The California Supreme Court, where the case eventually was heard, ruled that "therapists cannot escape liability merely because Tatiana herself was not their patient. When a therapist determines . . . that his patient presents a serious danger of violence to another, he incurs an obligation to use reasonable care to protect the intended victim against such danger." The steps the court included were warning the intended victim, warning others who would apprise the intended victim of the danger, and warning the police. The court went on to state:

> We recognize the public interest in supporting effective treatment of mental illness and in protecting the rights of patients to privacy, and the consequent public importance of

safeguarding the confidential character of psychotherapeutic communication. Against this interest, however, we must weigh the public interest in safety from violent assault.

The opinion closed with the following:

> We conclude that the public policy favoring protection of the confidential character of patient-psychotherapist communication must yield to the extent to which disclosure is essential to avert danger to others. The protective privilege ends where the public peril begins.

This case established a "duty to protect" for individuals who treat patients who appear to present an imminent danger to an identifiable person or persons. The ruling appears to apply mainly to therapists, but here the waters are muddy. Human service professionals in all states have taken the position that if such circumstances were to occur in the course of their work, the courts would find them negligent if they had not exercised the precautions laid out in the Tarasoff case. Some states actually have statutory or binding case law that establishes the duty to warn, while others do not. Regardless, you must assume that the courts would find you or your agency negligent if you failed to take the precautions outlined in the Tarasoff ruling. It is unlikely that you would be excused from liability because you are a case manager, and not a therapist.

Rarely would you make the decision to warn alone. If you believe a client poses an imminent danger to another identifiable person or persons, you must take the matter up at once with your supervisor. If your supervisor is not available for consultation and you believe you cannot wait, notify the police.

Assignment

Find out what laws exist, if any, in your state regarding your duty to warn. If there are no laws on the books, what is common legal opinion regarding the duty to warn?

Mandated Reporting

All states have laws requiring professionals to report the abuse and neglect of children. In some states, there are laws requiring the human service worker to report elder abuse. Professionals who must report abuse and neglect under the law are called "mandated reporters." The definition of child abuse and elder abuse varies from state to state. The laws in each state stipulate who is a mandated reporter, and there are variations among the states in regard to which professionals are considered mandated to report.

Even in states where there is no mandate to report elder abuse, there may be protective services for the elderly where you can report suspected abuse of an older person. You have an ethical responsibility not to ignore abuse of this type, regardless of the law. It is your responsibility to protect the clients, particularly individuals who cannot protect themselves.

Assignment

Find out your state's definition of child abuse. Learn which professionals in your state are considered mandated reporters of child abuse. What are the laws in your state regarding elder abuse?

DIAGNOSTIC LABELING

Agencies that rely on a diagnosis in order to be paid for service by a third-party payor (such as insurance companies, Medicare, and Medicaid) need to inform the client of that fact. Clients rarely understand that labels are used in this way, and most clients do not know the labels. They are rarely clear about the fact that the information will be passed on to their insurance companies. Clients should be informed of this. The client can then decide whether to continue to receive services from the agency. A client may elect to leave the agency or to pay for the services himself, without involving his insurance company, as a means of ensuring his privacy. Unless he is informed, he will not know he has these choices.

Another point about diagnosing clients is that practitioners use the categories of illness to know which treatment to use and how to develop the most effective treatment plan. Much research has been done to link the best treatments with each of the diagnostic categories. Clients will appreciate the need for a diagnostic label if they understand this. What may appear to clients as simple respect, kindness, good communication, or personal support on the part of their therapists may actually be the use of well-developed treatment modes.

INVOLUNTARY COMMITMENT

Generally an involuntary commitment occurs to a facility that specializes in inpatient mental health care. It could be a unit in a general hospital in the community where the client lives, a private psychiatric hospital, or, in some cases, a partial hospitalization program where the client receives treatment during that portion of the day that he or she is most at risk.

Patients have a right to expect the least restrictive form of treatment. If they need hospitalization, but not a locked ward, they should not be locked up twenty-four hours a day. If they can get the care they need in a partial hospitalization program, they should not have to go into the hospital. In talking about the movement to deinstitutionalize mental patients, Bednar et al. (1991) wrote, "[T]reatment should be no more harsh, hazardous, or intrusive than necessary to achieve therapeutic aims and to protect clients and others from physical harm."

The courts take seriously their responsibility to commit individuals in need of psychiatric care who are unable to obtain it because of a current severe impairment. In making the commitment, the courts make it clear that the purpose is treatment, and not punishment for behavior. For that reason, court commitment proceedings are often less formal and more pleasant than criminal proceedings. Students may observe these proceedings if they choose, as the proceedings are public. If you are actually involved in a commitment procedure, be sure to document all the steps you take in order to protect yourself from liability. The criteria for committing someone against his or her will are as follows:

1. The person poses a danger to him/herself or to others, *and possibly one or more of the following:*
2. The person has a severe mental illness or a mental illness that is currently acute.
3. The person is unable to function in occupational, social, or personal areas. The impairment is severe enough that the person cannot care adequately for him/herself.
4. The person has refused to sign a voluntary commitment, committing him/herself for treatment, so that an involuntary commitment is the last resort; or the person is incapable of signing such a commitment or of choosing treatment for him/herself.

5. The person can be treated once he or she is committed; that is, there are known treatments and medications that can relieve the acute condition the person is experiencing at present.

6. The commitment adheres to the criteria of the least restrictive treatment setting.

Assignment

Find out what the commitment laws are in your state. Look at the various types of voluntary and involuntary commitments. Find out under what circumstances the client can leave a facility when voluntarily committed. Find out what constitutes due process in your state for those being committed involuntarily.

ETHICAL RESPONSIBILITY

Responsibility for the client's welfare while the client is in your program is yours. The consumer views you as an authority. No matter how inexperienced you feel, when clients work with you, they will see you as the person with all the answers. For this reason, you will have considerable influence over what your clients decide to do. It is important to keep their needs at the forefront of your planning and delivery of services.

Burdening the Client with Your Problems

Sometimes people in human services use clients to meet their own needs. You could, for instance, burden the client with your own problems. You might say things like "Oh, that happened to me too" or "Wait'll you hear what happened to me!" You might have had a bad day and want to talk to someone about it, and so you tell your client all about it, as Kent did with his client Lucy.

Meeting Your Needs

Do not ask the client to do something that meets your needs or is not in the best interest of your client. Because of the influence you have with this person, it is easy to influence a client to do something that is beneficial to you. You might get the client involved with a friend of yours who sells insurance, or you might ask the client to go on television with you or to do an interview for the paper. There might be some payoff for you, but because the client could have considerable difficulty saying no to you, it is imperative that you never place your client in this situation. Often the media wants to interview a person with schizophrenia or a recovering alcoholic. Inform members of the media that they will have to locate their own interviewees.

Insisting on Your Solutions

You may have a need to look efficient, innovative, or particularly therapeutic, and so you might try to give your clients solutions to their problems. You may have had a similar problem at one time and feel there is only one good way to resolve it—the way you used to resolve it, a way your client can use without experiencing the hard knocks you took figuring it out. You might be tempted to lecture, to discuss your situation and how it compares with theirs, or to warn your clients. None of these actions will help your clients to grow by finding their own solutions.

Another way you can make a client do what you want her to do is to treat her rudely if she fails to use your solutions or to move quickly enough toward a solution. Being rude is not the same thing as being firm. You can set limits, but it is inappropriate to treat a client brusquely for not improving or for not taking what you see as healthy measures.

Exploiting Dependency

Clients are naturally vulnerable. They come to you at a time in their lives when they are hurt, upset, and disorganized—a time when it is easy to come to rely on another person. You are in a position to exploit this vulnerability by maintaining the client in a dependent position long after such dependency is useful for the client. For example, you might enjoy having clients call you about the details and decisions of their lives. It might make you feel important or needed. You might encourage them to lean on you for assistance in matters they could manage themselves. Be very careful not to allow this sort of relationship to develop.

COMPETENCE

A significant characteristic of professionals is their ability to clearly know their limitations. Ethical professionals do not try to do work for which they have not been trained. Recognizing the limits of one's training and experience is very important.

This means that you will be aware of areas where you could use some help or direction and that you will seek assistance when you need assistance. You will ask those who have more experience or education to assist you, rather than attempting to do work for which you are not qualified.

In addition, seek additional training throughout your career. Most certification and licensing programs require that individuals obtain further training on a yearly basis. Even if you are not part of such a program, you have an ethical responsibility to increase your skills, knowledge, and understanding of the field in which you work.

PROFESSIONAL RESPONSIBILITY

Finally, remember that you represent an agency and that it is your responsibility to establish a relationship with your client that is befitting of the agency. This will affect your relationship with the client in two ways.

First, know the parameters of your agency and operate within them. If you work for an agency that gives out food and fuel to the poor, do not attempt to do mental health counseling. If you are a case manager in a drug and alcohol unit, do not attempt to arrange foster care for one of your client's children except through the agency designated to handle that. If you work in a shelter for battered women, do not try to do drug rehabilitation. When a client needs services that fall outside the particular focus of your agency, make a referral to another agency that can best handle the problem.

The second way your relationship with the client is affected is related to dual relationships. Remain professional. Limit your contact to the focus of your agency and to the focus of that particular client's problems. Do not invite people home to dinner, take them home with you for the night, or become socially involved with them because you feel sorry for them.

Perhaps you are in the ER giving assistance to a woman who has been robbed in her home. You are working for a rape crisis center. It is late. The woman you are interviewing is terrified of going home. You call crisis intervention to arrange for temporary lodging, but they are currently out of the office on a call and will have to "get back to you." The woman has tried unsuccessfully to reach two family members but has reached only their answering machines. Finally, in desperation, you take the woman home with you. You would rather do this than sit in the ER all night because you have to be at a meeting in the morning. The woman is educated and seems very pleasant and refined. She goes home with you, spends the night at your house, and returns with you to the agency in the morning, where you help her obtain temporary housing and see that she gets safely to work.

Two months later you receive a call in the middle of the night. It is the same client. She has had a fight with her boyfriend, and now she wants to stay with you. You tell her she cannot do that, and she becomes hysterical. In the next several weeks, she appears at your house several times, asking to stay with you. Always there is some reason why she cannot stay where she is currently living. She knows your phone number, so she calls frequently. You have to be very firm in order to set limits, and even then it is hard to do.

This story is not all that unusual for workers who have taken someone home with them. Rarely is a person who is hurting and vulnerable able to reestablish a professional relationship with such a worker, complete with boundaries and limits, once the worker has extended this kind of friendship or kindness.

SUMMARY

This chapter is particularly important because it involves your ethical obligation to the client and outlines some legal concepts you must follow to protect your client and yourself. The primary issue is always the welfare of the client. That must come before all other considerations. When we choose this line of work, we deliberately choose to work with vulnerable people who cannot be expected to protect themselves or to know their rights. It becomes our responsibility to see that clients are well protected and are treated or given service under the highest ethical standards.

ETHICS EXERCISE

Instructions: The hypothetical practice situations that follow are designed to stimulate thinking and discussion on the issue of confidentiality. Each situation is followed by a multiple-choice list of possible responses you might make. Choose the response that you consider the best.

1. Paula is a 17-year-old client in the daytime partial hospitalization program. Her mother phoned and requested to know Paula's psychiatric diagnosis so that she could inform the family's physician who is treating Paula for diabetes. You should:

    a. Advise the mother of the diagnosis and the name of the psychiatrist who made the diagnosis.
    b. Call the family physician directly and advise him of the diagnosis.
    c. Ask Paula to sign a release of information form giving consent for the physician and/or the mother to be advised of her diagnosis.
    d. Refuse to release the information at all.

2. A client requests a copy of his current treatment plan. You should:

    a. Have the client put the request in writing and discuss the issue with the treatment team.
    b. Make a copy of the current treatment plan and give it to the client.
    c. Discuss the treatment plan and give it to the client.
    d. Refer the client to the attending psychiatric physician.

3. A 13-year-old client requests that his school counselor be sent a copy of his initial interview and discharge summary. The client signs a release of information form, documenting his written consent for the information to be transmitted. You should:

    a. Forward the material to the school counselor.
    b. Give the information to the client who can deliver it to the school counselor.
    c. Have the medical records department forward the information to the school counselor.
    d. Refuse to release the information until a parent cosigns the release of information form.

4. Mary Smith is a depressed elderly woman who was admitted to Polyclinic Hospital due to severe back pain. She was advised she might need surgery to correct the problem. You are her case manager at the Office of Aging, where she calls to say she is considering suicide. The constant back pain has made her feel like "just giving up." Mary is currently at home, awaiting a surgery date. You know Mary has a supply of pain pills, and she says she wants to take all the pills. You feel there is a substantial risk of Mary following through on her threat. You should:

    a. Contact the Polyclinic orthopedic staff who are currently seeing Mary in outpatient clinic.
    b. Maintain frequent contact with Mary, but respect her wishes to keep her suicide plans confidential.
    c. After discussing with Mary what you are about to do, contact crisis intervention.
    d. Advise the city police department of Mary's suicide plans.

5. Bill Jones is a client who has been in alcohol treatment programs at your facility. He is currently depressed about his pending divorce and current marital separation. He has signed a release of information form for you to share information with his priest, who is counseling Bill about his religious conflicts regarding the divorce. A man calls you claiming to be Bill's priest and requesting information on Bill's current state of mind. You have never actually spoken with Bill's priest, and you think this might actually be Bill's wife's attorney calling. You should:

    a. Verify the identity of the caller before giving any information.
    b. Refer the call to Bill.
    c. Insist upon meeting with the priest in person.
    d. Refuse to share any information with the caller.
    e. Get his number and call him back.

6. Patty is completing a student internship for her associate's degree in the therapeutic activities program. She asks to review the medical records of the clients who were just in her projects group. You are supervising Patty. You should:

    a. Advise Patty that the records are confidential and may not be inspected by students.
    b. Make certain that Patty is well trained in the policies and procedures relating to confidential information, and only then allow her access to the medical records.
    c. Permit Patty free access to the records because she is like part of the staff.
    d. Obtain written consent from each client for Patty to review the records.

7. Jerry was a client who improved and was discharged two years ago. You receive a call from the National Can Company. The caller explains that Jerry has applied for a job and that the company would like to hire him. Jerry told them he was in treatment two years ago and was discharged after showing considerable improvement. The company wishes to confirm the fact that Jerry did indeed complete the program as he claims. You should:

    a. Be very careful not to reveal whether or not Jerry was ever in treatment until a signed release form is received.
    b. Confirm that Jerry completed his treatment but make no comment on progress or condition at discharge.
    c. Explain, on the phone, that Jerry did successfully complete treatment and request that the company forward a release of information form to you in the next mail.
    d. Make no comment and hang up quickly.

8. Clark is currently enrolled in treatment, and you are his case manager. He asks you if he may read his medical record. You should:

    a. Ask Clark to put the request in writing, and assist Clark in completing the written request if he seems to have limited skills in reading and writing.
    b. Present Clark's request to Clark's treatment team.
    c. If the treatment team concludes that it will not harm Clark to review his record, allow Clark to read it in the presence of a therapist (after deleting information from sources that asked to remain anonymous).
    d. Decide with the treatment team who will assist Clark in reading and understanding his record. Then follow through by allowing Clark to review his record with that person.

ETHICALLY, WHAT WENT WRONG?

Instructions: The following hypothetical practice situations are designed to help you apply what you have learned in this chapter. For each situation, decide what was done in the situation that was unethical.

1. Jennifer had a long day and was trying to get out of the office before 5:00 P.M. She had one more client to see. Dr. Adams had asked Jennifer to give the client, a young man recently diagnosed with schizophrenia, a prescription for new medications. Jennifer had her coat on when she handed the prescriptions to the client in the waiting room. The client wanted to know what the medication was and why his prescription was being changed. "Will there be any side effects?" he asked Jennifer.

She replied hurriedly, "Oh, no. Dr. Adams says just take this until he sees you next time."

2. Carl is uncomfortable around gay men. A client of his is gay and has just broken up with his lover. This client, a 42-year-old man who had been in a long-term relationship, is devastated and is in tears in Carl's office. Because the client has suffered from severe depression in the past, Carl is attempting to have him evaluated by the therapist this afternoon. In the meantime, the client is weeping and threatening to take his life. Carl is particularly uncomfortable with this man's tears and believes this is all a drama. Carl says, "Oh, c'mon now. Let's get a grip. You can't sit in here all afternoon carrying on. Here. Take some tissue and go out in the waiting room until Dr. Paul can see you."

3. Elizabeth visited in the home of an elderly man and got him to sign a release of information form so she could process an application to the county nursing home. In the man's records were references to the fact that many years ago as a teenager he was convicted of shooting a man in a bar fight, a crime for which he served two years in prison. She knows the people at the home will be titillated over this little tidbit of information, especially her friend Rhoda, who does the intakes. Even though she knows this is not part of the home's evaluation, that the client has led an exemplary life since that time, and that nursing staff at the home might take it out of context, she releases the information anyway, based on her client's signature on the release form. She and Rhoda have a good laugh about it the next day.

4. Jim is doing an intake with a man who claims he is depressed. He tells Jim that ever since his wife left he has had trouble concentrating and waking up in the morning. He talks about how lonely it is at home, how much he misses his children, how he is tempted to drink in the evenings, and how little he has to look forward to. Jim nods. He understands. "Yes, my wife left last month too. I know just what you mean. I get to feeling like, well like there isn't as much meaning. I never knew the kids were that important to me. I guess they were. On Saturdays I used to do things with my son and I still get him every other weekend, but it's not quite the same thing, is it?" "No," the other man responds, "I was thinking . . . " "Well, I do a lot of thinking, too," Jim interrupts him. "I think about what I could have done differently and if it was my fault. Don't you think these women would see that it's hard, too hard I think, to raise kids alone?" The conversation continues in this vein until the end of the interview.

5. Carmen is supposed to see her small caseload of persistently mentally ill individuals at least twice a week. Lately with school and her mother's death, she has not really seen them that often. She has checked in with them on the phone, but she has also used time when she was out seeing clients to do errands at the library or empty her mother's home. Now one of her clients is in court after committing a crime. The client and the lawyer agree that the client might be able to use his mental health status as a reason for committing the crime, and they ask to introduce the case record as evidence in the court proceedings. Fearing that it will be discovered how little supervision and attention she had given her client and knowing that she could ultimately be blamed for the fact that her client committed the crime while under her somewhat irresponsible care, Carmen invokes the concept of privileged communication to keep from giving the file to the court.

6. Ted is in a clinic with his elderly client for a routine blood workup, which they do every other month. He notices the client is bruised on the face and arms. For a while he makes small talk with her, and then he asks her about the bruises. She is somewhat evasive but indicates, "They weren't the result of no fall!" Without explicitly blaming her daughter and son-in-law, with whom she lives, she makes it quite clear that the bruises are not the result of an accident. After the blood test, during which neither the doctor, who sees her briefly, nor the technician make any mention of the bruises, Ted takes his client home. He toys with the idea of reporting the bruises to the protective services at the county Office of Aging but decides not to. He bases his decision on the fact that the law does not specifically require him to do so, that it would be hard and take a lot of time to have to place this client in another living arrangement, and that the daughter seems like a very nice person who Ted does not feel like stirring up over an uncomfortable situation.

7. Kitty has a whole list of things to do today and doubts she can get it all done. She hates the way there are always things left to do at the end of the day. It just seems that no matter how hard she works, something new comes up that she cannot complete. One of her clients has told her on the phone that she wants to sign a release of information form for her lawyer. Kitty has the form ready for the time when this client will be coming in at the end of the week. Today a man calls and says he is the client's lawyer and he needs just two dates to help him file a brief with the court on the client's behalf. Kitty gives him the two dates and hurries to the next thing on her list.

8. Jorge is mad at lunch in the staff room this noon. He spent one morning taking a meticulous social history from a new client. The client, a man in his twenties, was

pleasant and helpful. He seemed to genuinely want the assistance of the agency and to like Jorge. Two more interviews followed to set up services, and the client signed a release of information form for Jorge to meet with the client's physician. Jorge cannot understand why the client never mentioned the fact that he is HIV+. This Jorge found out in the conference with the client's physician some weeks later. "How do these clients think I am going to help them if they don't tell the whole story?" Jorge fumed. "They come in here and want my help and then withhold information from me. They leave me in the dark. I don't know what's going on, and then they think I'm going to be able to help them."

ETHICAL STANDARDS OF HUMAN SERVICE PROFESSIONALS

National Organization for Human Service Education Council for Standards in Human Service Education*

Preamble

Human services is a profession developing in response to and in anticipation of the direction of human needs and human problems in the late twentieth century. Characterized particularly by an appreciation of human beings in all their diversity, human services offers assistance to its clients within the context of their community and environment. Human service professionals, regardless of whether they are students, faculty, or practitioners, promote and encourage the unique values and characteristics of human services. In so doing human service professionals uphold the integrity and ethics of the profession, partake in constructive criticism of the profession, promote the client and community well-being, and enhance their own professional growth.

The ethical guidelines presented are a set of standards of conduct which the human service professional considers in ethical and professional decision making. It is hoped that these guidelines will be of assistance when the human service professional is challenged by difficult ethical dilemmas. Although ethical codes are not legal documents, they may be used to assist the adjudication of issues related to ethical human service behavior.

Human service professionals function in many ways and carry out many roles. They enter into professional-client relationships with individuals, families, groups, and communities who are all referred to as "clients" in these standards. Among their roles are caregiver, case manager, broker, teacher/educator, behavior changer, consultant, outreach professional, mobilizer, advocate, community planner, community change organizer, evaluator, and administrator.[1] The following standards are written with these multi-faceted roles in mind.

---

1. Southern Regional Education Board. (1967). *Roles and Functions from Mental Health Workers: A Report of a Symposium.* Atlanta, GA: Community Mental Health Worker Project.
*Reprinted with permission.

The Human Service Professional's Responsibility to Clients

Statement 1   Human service professionals negotiate with clients the purpose, goals, and nature of the helping relationship prior to its onset, as well as inform clients of the limitations of the proposed relationship.

Statement 2   Human service professionals respect the integrity and welfare of the client at all times. Each client is treated with respect, acceptance, and dignity.

Statement 3   Human service professionals protect the client's right to privacy and confidentiality except when such confidentiality would cause harm to the client or others, when agency guidelines state otherwise, or under other stated conditions (e.g., local, state, or federal laws). Professionals inform clients of the limits of confidentiality prior to the onset of the helping relationship.

Statement 4   If it is suspected that danger or harm may occur to the client or to others as a result of a client's behavior, the human service professional acts in an appropriate and professional manner to protect the safety of those individuals. This may involve seeking consultation, supervision, and/or breaking confidentiality of the relationship.

Statement 5   Human service professionals protect the integrity, safety, and security of client records. All written client information that is shared with other professionals, except in the course of professional supervision, must have the client's prior written consent.

Statement 6   Human service professionals are aware that in their relationships with clients power and status are unequal. Therefore, they recognize that dual or multiple relationships may increase the risk of harm to, or exploitation of, clients, and may impair their professional judgment. However, in some communities and situations it may not be feasible to avoid social or other nonprofessional contact with clients. Human service professionals support the trust implicit in the helping relationship by avoiding dual relationships that may impair professional judgment, increase the risk of harm to clients, or lead to exploitation.

Statement 7   Sexual relationships with current clients are not considered to be in the best interest of the client and are prohibited. Sexual relationships with previous clients are considered dual relationships and are addressed in statement 6 (above).

Statement 8   The client's right to self-determination is protected by human service professionals. They recognize the client's right to receive or refuse services.

Statement 9   Human service professionals recognize and build on client strengths.

The Human Service Professional's Responsibility to the Community and Society

Statement 10   Human service professionals are aware of local, state, and federal laws. They advocate for change in regulations and statutes when such legislation conflicts with ethical guidelines and/or client rights. Where laws are harmful to individuals, groups, or communities, human service professionals consider the conflict between the values of obeying the law and the values of serving people and may decide to initiate social action.

Statement 11 Human service professionals keep informed about current social issues as they affect the client and the community. They share that information with clients, groups, and community as part of their work.

Statement 12 Human service professionals understand the complex interaction between individuals, their families, the communities in which they live, and society.

Statement 13 Human service professionals act as advocates in addressing unmet client and community needs, calling attention to these needs, and assisting in planning and mobilizing to advocate for those needs at the local community level.

Statement 14 Human service professionals represent their qualifications to the public accurately.

Statement 15 Human service professionals describe the effectiveness of programs, treatments, and/or techniques accurately.

Statement 16 Human service professionals advocate for the rights of all members of society, particularly those who are members of minorities and groups at which discriminatory practices have historically been directed.

Statement 17 Human service professionals provide services without discrimination or preference based on age, ethnicity, culture, race, disability, gender, religion, sexual orientation, or socioeconomic status.

Statement 18 Human service professionals are knowledgeable about cultures and communities within which they practice. They are aware of multiculturalism in society and its impact on the community as well as individuals within the community. They respect individuals and groups, their cultures and beliefs.

Statement 19 Human service professionals are aware of their own cultural backgrounds, beliefs, and values, recognizing the potential for impact on their relationships with others.

Statement 20 Human service professionals are aware of sociopolitical issues that differentially affect clients from diverse backgrounds.

Statement 21 Human service professionals seek training, experience, education, and supervision necessary to ensure their effectiveness in working with culturally diverse client populations.

The Human Service Professional's Responsibility to Colleagues

Statement 22 Human service professionals avoid duplicating another professional's helping relationship with a client. They consult with other professionals who are assisting the client in a different type of relationship when it is in the best interest of the client to do so.

Statement 23 When a human service professional has a conflict with a colleague, he or she first seeks out the colleague in an attempt to manage the problem. If necessary, the professional then seeks the assistance of supervisors, consultants, or other professionals in efforts to manage the problem.

Statement 24   Human service professionals respond appropriately to unethical behavior of colleagues. Usually this means initially talking directly with the colleague and, if no resolution is forthcoming, reporting the colleague's behavior to supervisory or administrative staff and/or to the professional organization(s) to which the colleague belongs.

Statement 25   All consultations between human service professionals are kept confidential unless to do so would result in harm to clients or communities.

The Human Service Professional's Responsibility to the Profession

Statement 26   Human service professionals know the limit and scope of their professional knowledge and offer services only within their knowledge and skill base.

Statement 27   Human service professionals seek appropriate consultation and supervision to assist in decision making when there are legal, ethical, or other dilemmas.

Statement 28   Human service professionals act with integrity, honesty, genuineness, and objectivity.

Statement 29   Human service professionals promote cooperation among related disciplines (e.g., psychology, counseling, social work, nursing, family and consumer sciences, medicine, education) to foster professional growth and interests within the various fields.

Statement 30   Human service professionals promote the continuing development of their profession. They encourage membership in professional associations, support research endeavors, foster educational advancement, advocate for appropriate legislative actions, and participate in other related professional activities.

Statement 31   Human service professionals continually seek out new and effective approaches to enhance their professional abilities.

The Human Service Professional's Responsibility to Employers

Statement 32   Human service professionals adhere to commitments made to their employers.

Statement 33   Human service professionals participate in efforts to establish and maintain employment conditions which are conducive to high quality client services. They assist in evaluating the effectiveness of the agency through reliable and valid assessment measures.

Statement 34   When a conflict arises between fulfilling the responsibility to the employer and the responsibility to the client, human service professionals advise both of the conflict and work conjointly with all involved to manage the conflict.

The Human Service Professional's Responsibility to Self

Statement 35   Human service professionals strive to personify those characteristics typically associated with the profession (e.g., accountability, respect for others, genuineness, empathy, pragmatism).

Statement 36 Human service professionals foster self-awareness and personal growth in themselves. They recognize that when professionals are aware of their own values, attitudes, cultural background, and personal needs, the process of helping others is less likely to be negatively impacted by those factors.

Statement 37 Human service professionals recognize a commitment to lifelong learning and continually upgrade knowledge and skills to serve the population better.

NASW CODE OF ETHICS*
Overview

The *NASW Code of Ethics* is intended to serve as a guide to the everyday professional conduct of social workers. This *Code* includes four sections. The first section, "Preamble," summarizes the social work profession's mission and core values. The second section, "Purpose of the *NASW Code of Ethics*," provides an overview of the *Code's* main functions and a brief guide for dealing with ethical issues or dilemmas in social work practice. The third section, "Ethical Principles," presents broad ethical principles, based on social work's core values, that inform social work practice. The final section, "Ethical Standards," includes specific ethical standards to guide social workers' conduct and to provide a basis for adjudication.

Preamble

The primary mission of the social work profession is to enhance human well-being and help meet the basic human needs of all people, with particular attention to the needs and empowerment of people who are vulnerable, oppressed, and living in poverty. A historic and defining feature of social work is the profession's focus on individual well-being in a social context and the well-being of society. Fundamental to social work is attention to the environmental forces that create, contribute to, and address problems in living.

Social workers promote social justice and social change with and on behalf of clients. "Clients" is used inclusively to refer to individuals, families, groups, organizations, and communities. Social workers are sensitive to cultural and ethnic diversity and strive to end discrimination, oppression, poverty, and other forms of social injustice. These activities may be in the form of direct practice, community organizing, supervision, consultation, administration, advocacy, social and political action, policy development and implementation, education, and research and evaluation. Social workers seek to enhance the capacity of people to address their own needs. Social workers also seek to promote the responsiveness of organizations, communities, and other social institutions to individuals' needs and social problems.

The mission of the social work profession is rooted in a set of core values. These core values, embraced by social workers throughout the profession's history, are the foundation of social work's unique purpose and perspective:

service
social justice
dignity and worth of the person
importance of human relationships
integrity
competence.

This constellation of core values reflects what is unique to the social work profession. Core values, and the principles that flow from them, must be balanced within the context and complexity of the human experience.

Purpose of the NASW Code of Ethics

Professional ethics are at the core of social work. The profession has an obligation to articulate its basic values, ethical principles, and ethical standards. The *NASW Code of Ethics* sets forth these values, principles, and standards to guide social workers' conduct. The *Code* is relevant to all social workers and social work students, regardless of their professional functions, the settings in which they work, or the populations they serve.

The *NASW Code of Ethics* serves six purposes:

1. The *Code* identifies core values on which social work's mission is based.
2. The *Code* summarizes broad ethical principles that reflect the profession's core values and establishes a set of specific ethical standards that should be used to guide social work practice.
3. The *Code* is designed to help social workers identify relevant considerations when professional obligations conflict or ethical uncertainties arise.
4. The *Code* provides ethical standards to which the general public can hold the social work profession accountable.
5. The *Code* socializes practitioners new to the field to social work's mission, values, ethical principles, and ethical standards.
6. The *Code* articulates standards that the social work profession itself can use to assess whether social workers have engaged in unethical conduct. NASW has formal procedures to adjudicate ethics complaints filed against its members. In subscribing to this *Code,* social workers are required to cooperate in its implementation, participate in NASW adjudication proceedings, and abide by any NASW disciplinary rulings or sanctions based on it.

The *Code* offers a set of values, principles, and standards to guide decision making and conduct when ethical issues arise. It does not provide a set of rules that prescribe how social workers should act in all situations. Specific applications of the *Code* must take into account the context in which it is being considered and the possibility of conflicts among the *Code*'s values, principles, and standards. Ethical responsibilities flow from all human relationships, from the personal and familial to the social and professional.

Further, the *NASW Code of Ethics* does not specify which values, principles, and standards are most important and ought to outweigh others in instances when they conflict. Reasonable differences of opinion can and do exist among social workers with respect to the ways in which values, ethical principles, and ethical standards should be rank ordered when they conflict. Ethical decision making in a given situation must apply the informed judgment of the individual social worker and should also consider how the issues would be judged in a peer review process where the ethical standards of the profession would be applied.

Ethical decision making is a process. There are many instances in social work where simple answers are not available to resolve complex ethical issues. Social workers should take into consideration all the values, principles, and standards in this *Code* that are relevant to any situation in which ethical judgment is warranted. Social workers' decisions and actions should be consistent with the spirit as well as the letter of this *Code.*

In addition to this *Code*, there are many other sources of information about ethical thinking that may be useful. Social workers should consider ethical theory and principles generally, social work theory and research, laws, regulations, agency policies, and other relevant codes of ethics, recognizing that among codes of ethics social workers should consider the *NASW Code of Ethics* as their primary source. Social workers also should be aware of the impact on ethical decision making of their clients' and their own personal values and cultural and religious beliefs and practices. They should be aware of any conflicts between personal and professional values and deal with them responsibly. For additional guidance social workers should consult the relevant literature on professional ethics and ethical decision making and seek appropriate consultation when faced with ethical dilemmas. This may involve consultation with an agency-based or social work organization's ethics committee, a regulatory body, knowledgeable colleagues, supervisors, or legal counsel.

Instances may arise when social workers' ethical obligations conflict with agency policies or relevant laws or regulations. When such conflicts occur, social workers must make a responsible effort to resolve the conflict in a manner that is consistent with the values, principles, and standards expressed in this *Code*. If a reasonable resolution of the conflict does not appear possible, social workers should seek proper consultation before making a decision.

The *NASW Code of Ethics* is to be used by NASW and by individuals, agencies, organizations, and bodies (such as licensing and regulatory boards, professional liability insurance providers, courts of law, agency boards of directors, government agencies, and other professional groups) that choose to adopt it or use it as a frame of reference. Violation of standards in this *Code* does not automatically imply legal liability or violation of the law. Such determination can only be made in the context of legal and judicial proceedings. Alleged violations of the *Code* would be subject to a peer review process. Such processes are generally separate from legal or administrative procedures and insulated from legal review or proceedings to allow the profession to counsel and discipline its own members.

A code of ethics cannot guarantee ethical behavior. Moreover, a code of ethics cannot resolve all ethical issues or disputes or capture the richness and complexity involved in striving to make responsible choices within a moral community. Rather, a code of ethics sets forth values, ethical principles, and ethical standards to which professionals aspire and by which their actions can be judged. Social workers' ethical behavior should result from their personal commitment to engage in ethical practice. The *NASW Code of Ethics* reflects the commitment of all social workers to uphold the profession's values and to act ethically. Principles and standards must be applied by individuals of good character who discern moral questions and, in good faith, seek to make reliable ethical judgments.

Ethical Principles

The following broad ethical principles are based on social work's core values of service, social justice, dignity and worth of the person, importance of human relationships, integrity, and competence. These principles set forth ideals to which all social workers should aspire.

Value: *Service*

Ethical Principle: *Social workers' primary goal is to help people in need and to address social problems.*

Social workers elevate service to others above self-interest. Social workers draw on their knowledge, values, and skills to help people in need and to address social problems.

Social workers are encouraged to volunteer some portion of their professional skills with no expectation of significant financial return (pro bono service).

Value: *Social Justice*

Ethical Principle: *Social workers challenge social injustice.*

Social workers pursue social change, particularly with and on behalf of vulnerable and oppressed individuals and groups of people. Social workers' social change efforts are focused primarily on issues of poverty, unemployment, discrimination, and other forms of social injustice. These activities seek to promote sensitivity to and knowledge about oppression and cultural and ethnic diversity. Social workers strive to ensure access to needed information, services, and resources; equality of opportunity; and meaningful participation in decision making for all people.

Value: *Dignity and Worth of the Person*

Ethical Principle: *Social workers respect the inherent dignity and worth of the person.*

Social workers treat each person in a caring and respectful fashion, mindful of individual differences and cultural and ethnic diversity. Social workers promote clients' socially responsible self-determination. Social workers seek to enhance clients' capacity and opportunity to change and to address their own needs. Social workers are cognizant of their dual responsibility to clients and to the broader society. They seek to resolve conflicts between clients' interests and the broader society's interests in a socially responsible manner consistent with the values, ethical principles, and ethical standards of the profession.

Value: *Importance of Human Relationships*

Ethical Principle: *Social workers recognize the central importance of human relationships.*

Social workers understand that relationships between and among people are an important vehicle for change. Social workers engaged people as partners in the helping process. Social workers seek to strengthen relationships among people in a purposeful effort to promote, restore, maintain, and enhance the well-being of individuals, families, social groups, organizations, and communities.

Value: *Integrity*

Ethical Principle: *Social workers behave in a trustworthy manner.*

Social workers are continually aware of the profession's mission, values, ethical principles, and ethical standards and practice in a manner consistent with them. Social workers act honestly and responsibly and promote ethical practices on the part of the organizations with which they are affiliated.

Value: *Competence*

Ethical Principle: *Social workers practice within their areas of competence and develop and enhance their professional expertise.*

Social workers continually strive to increase their professional knowledge and skills and to apply them in practice. Social workers should aspire to contribute to the knowledge base of the profession.

Ethical Standards

The following ethical standards are relevant to the professional activities of all social workers. These standards concern (1) social workers' ethical responsibilities to clients, (2) social workers' ethical responsibilities to colleagues, (3) social workers' ethical responsibilities in practice settings, (4) social workers' ethical responsibilities as professionals, (5) social workers' ethical responsibilities to the social work profession, and (6) social workers' ethical responsibilities to the broader society.

Some of the standards that follow are enforceable guidelines for professional conduct, and some are aspirational. The extent to which each standard is enforceable is a matter of professional judgment to be exercised by those responsible for reviewing alleged violations of ethical standards.

1. Social Workers' Ethical Responsibilities to Clients

1.01 Commitments to Clients   Social workers' primary responsibility is to promote the well-being of clients. In general, clients' interests are primary. However, social workers' responsibility to the larger society or specific legal obligations may on limited occasions supersede the loyalty owed clients, and clients should be so advised. (Examples include when a social worker is required by law to report that a client has abused a child or has threatened to harm self or others.)

1.02 Self-Determination   Social workers respect and promote the right of clients to self-determination and assist clients in their efforts to identify and clarify their goals. Social workers may limit clients' right to self-determination when, in the social workers' professional judgment, clients' actions or potential actions pose a serious, foreseeable, and imminent risk to themselves or others.

1.03 Informed Consent   (a) Social workers should provide services to clients only in the context of a professional relationship based, when appropriate, on valid informed consent. Social workers should use clear and understandable language to inform clients of the purpose of the services, risks related to the services, limits to services because of the requirements of a third-party payer, relevant costs, reasonable alternatives, clients' right to refuse or withdraw consent, and the time frame covered by the consent. Social workers should provide clients with an opportunity to ask questions.

(b) In instances when clients are not literate or have difficulty understanding the primary language used in the practice setting, social workers should take steps to ensure clients' comprehension. This may include providing clients with a detailed verbal explanation or arranging for a qualified interpreter or translator whenever possible.

(c) In instances when clients lack the capacity to provide informed consent, social workers should protect clients' interests by seeking permission from an appropriate third party, informing clients consistent with the clients' level of understanding. In such instances social workers should seek to ensure that the third party acts in a manner consistent with clients' wishes and interests. Social workers should take reasonable steps to enhance such clients' ability to give informed consent.

(d) In instances when clients are receiving services involuntarily, social workers should provide information about the nature and extent of services and about the extent of clients' right to refuse service.

(e) Social workers who provide services via electronic media (such as computer, telephone, radio, and television) should inform recipients of the limitations and risks associated with such services.

(f) Social workers should obtain clients' informed consent before audiotaping or videotaping clients or permitting observation of services to clients by a third party.

1.04 Competence   (a) Social workers should provide services and represent themselves as competent only within the boundaries of their education, training, license, certification, consultation received, supervised experience, or other relevant professional experience.

(b) Social workers should provide services in substantive areas or use intervention techniques or approaches that are new to them only after engaging in appropriate study, training, consultation, and supervision from people who are competent in those interventions or techniques.

(c) When generally recognized standards do not exist with respect to an emerging area of practice, social workers should exercise careful judgment and take responsible steps (including appropriate education, research, training, consultation, and supervision) to ensure the competence of their work and to protect clients from harm.

1.05 Cultural Competence and Social Diversity   (a) Social workers should understand culture and its function in human behavior and society, recognizing the strengths that exist in all cultures:

(b) Social workers should have a knowledge base of their clients' cultures and be able to demonstrate competence in the provision of services that are sensitive to clients' cultures and to differences among people and cultural groups.

(c) Social workers should obtain education about and seek to understand the nature of social diversity and oppression with respect to race, ethnicity, national origin, color, sex, sexual orientation, age, marital status, political belief, religion, and mental or physical disability.

1.06 Conflicts of Interest   (a) Social workers should be alert to and avoid conflicts of interest that interfere with the exercise of professional discretion and impartial judgment. Social workers should inform clients when a real or potential conflict of interest arises and take reasonable steps to resolve the issue in a manner that makes the clients' interests primary and protects clients' interests to the greatest extent possible. In some cases, protecting clients' interests may require termination of the professional relationship with proper referral of the client.

(b) Social workers should not take unfair advantage of any professional relationship or exploit others to further their personal, religious, political, or business interests.

(c) Social workers should not engage in dual or multiple relationships with clients or former clients in which there is a risk of exploitation or potential harm to the client. In instances when dual or multiple relationships are unavoidable, social workers should take steps to protect clients and are responsible for setting clear, appropriate, and culturally sensitive boundaries. (Dual or multiple relationships occur when social workers relate to clients in more than one relationship, whether professional, social, or business. Dual or multiple relationships can occur simultaneously or consecutively.)

(d) When social workers provide services to two or more people who have a relationship with each other (for example, couples, family members), social workers should clarify with all parties which individuals will be considered clients and the nature of social workers' professional obligations to the various individuals who are receiving services. Social workers who anticipate a conflict of interest among the individuals receiving services or who anticipate having to perform in potentially conflicting roles (for example, when a social worker is asked to testify in a child custody dispute or divorce proceedings involving clients) should clarify their role with the parties involved and take appropriate action to minimize any conflict of interest.

1.07 Privacy and Confidentiality   (a) Social workers should respect clients' right to privacy. Social workers should not solicit private information from clients unless it is essential to providing services or conducting social work evaluation or research. Once private information is shared, standards of confidentiality apply.

(b) Social workers may disclose confidential information when appropriate with valid consent from a client or a person legally authorized to consent on behalf of a client.

(c) Social workers should protect the confidentiality of all information obtained in the course of professional service, except for compelling professional reasons. The general expectation that social workers will keep information confidential does not apply when disclosure is necessary to prevent serious, foreseeable, and imminent harm to a client or other identifiable person or when laws or regulations require disclosure without a client's consent. In all instances, social workers should disclose the least amount of confidential information necessary to achieve the desired purpose; only information that is directly relevant to the purpose for which the disclosure is made should be revealed.

(d) Social workers should inform clients, to the extent possible, about the disclosure of confidential information and the potential consequences, when feasible before the disclosure is made. This applies whether social workers disclose confidential information on the basis of a legal requirement or client consent.

(e) Social workers should discuss with clients and other interested parties the nature of confidentiality and limitations of clients' right to confidentiality. Social workers should review with clients circumstances where confidential information may be requested and where disclosure of confidential information may be legally required. This discussion should occur as soon as possible in the social worker–client relationship and as needed throughout the course of the relationship.

(f) When social workers provide counseling services to families, couples, or groups, social workers should seek agreement among the parties involved concerning each individual's right to confidentiality and obligation to preserve the confidentiality of information shared by others. Social workers should inform participants in family, couples, or group counseling that social workers cannot guarantee that all participants will honor such agreements.

(g) Social workers should inform clients involved in family, couples, marital, or group counseling of the social worker's, employer's, and agency's policy concerning the social worker's disclosure of confidential information among the parties involved in the counseling.

(h) Social workers should not disclose confidential information to third-party payers unless clients have authorized such disclosure.

(i) Social workers should not discuss confidential information in any setting unless privacy can be ensured. Social workers should not discuss confidential information in public or semipublic areas such as hallways, waiting rooms, elevators, and restaurants.

(j) Social workers should protect the confidentiality of clients during legal proceedings to the extent permitted by law. When a court of law or other legally authorized body orders social workers to disclose confidential or privileged information without a client's consent and such disclosure could cause harm to the client, social workers should request that the court withdraw the order or limit the order as narrowly as possible or maintain the records under seal, unavailable for public inspection.

(k) Social workers should protect the confidentiality of clients when responding to requests from members of the media.

(l) Social workers should protect the confidentiality of clients' written and electronic records and other sensitive information. Social workers should take reasonable steps to ensure that clients' records are stored in a secure location and that clients' records are not available to others who are not authorized to have access.

(m) Social workers should take precautions to ensure and maintain the confidentiality of information transmitted to other parties through the use of computers, electronic mail, facsimile machines, telephones and telephone answering machines, and other electronic or computer technology. Disclosure of identifying information should be avoided whenever possible.

(n) Social workers should transfer or dispose of clients' records in a manner that protects clients' confidentiality and is consistent with state statutes governing records and social work licensure.

(o) Social workers should take reasonable precautions to protect client confidentiality in the event of the social worker's termination of practice, incapacitation, or death.

(p) Social workers should not disclose identifying information when discussing clients for teaching or training purposes unless the client has consented to disclosure of confidential information.

(q) Social workers should not disclose identifying information when discussing clients with consultants unless the client has consented to disclosure of confidential information or there is a compelling need for such disclosure.

(r) Social workers should protect the confidentiality of deceased clients consistent with the preceding standards.

1.08 Access to Records   (a) Social workers should provide clients with reasonable access to records concerning the clients. Social workers who are concerned that clients' access to their records could cause serious misunderstanding or harm to the client should provide assistance in interpreting the records and consultation with the client regarding the records. Social workers should limit clients' access to their records, or portions of their records, only in exceptional circumstances when there is compelling evidence that such access would cause serious harm to the client. Both clients' requests and the rationale for withholding some or all of the record should be documented in clients' files.

(b) When providing clients with access to their records, social workers should take steps to protect the confidentiality of other individuals identified or discussed in such records.

1.09 Sexual Relationships   (a) Social workers should under no circumstances engage in sexual activities or sexual contact with current clients, whether such contact is consensual or forced.

(b) Social workers should not engage in sexual activities or sexual contact with clients' relatives of other individuals with whom clients maintain a close personal relationship when there is a risk of exploitation or potential harm to the client. Sexual activity or sexual contact with clients' relatives or other individuals with whom clients maintain a personal relationship has the potential to be harmful to the client and may make it difficult for the social worker and client to maintain appropriate professional boundaries. Social workers—not their clients, their clients' relatives, or other individuals with whom the client maintains a personal relationship—assume the full burden for setting clear, appropriate, and culturally sensitive boundaries.

(c) Social workers should not engage in sexual activities or sexual contact with former clients because of the potential for harm to the client. If social workers engage in conduct contrary to this prohibition or claim that an exception to this prohibition is warranted because of extraordinary circumstances, it is social workers—not their clients—

who assume the full burden of demonstrating that the former client has not been exploited, coerced, or manipulated, intentionally or unintentionally.

(d) Social workers should not provide clinical services to individuals with whom they have had a prior sexual relationship. Providing clinical services to a former sexual partner has the potential to be harmful to the individual and is likely to make it difficult for the social worker and individual to maintain appropriate professional boundaries.

1.10 Physical Contact   Social workers should not engage in physical contact with clients when there is a possibility of psychological harm to the client as a result of the contact (such as cradling or caressing clients). Social workers who engage in appropriate physical contact with clients are responsible for setting clear, appropriate, and culturally sensitive boundaries that govern such physical contact.

1.11 Sexual Harassment   Social workers should not sexually harass clients. Sexual harassment includes sexual advances, sexual solicitation, requests for sexual favors, and other verbal or physical conduct of a sexual nature.

1.12 Derogatory Language   Social workers should not use derogatory language in their written or verbal communications to or about clients. Social workers should use accurate and respectful language in all communications to and about clients.

1.13 Payment for Services   (a) When setting fees, social workers should ensure that the fees are fair, reasonable, and commensurate with the services performed. Consideration should be given to clients' ability to pay.

(b) Social workers should avoid accepting goods or services from clients as payment for professional services. Bartering arrangements, particularly involving services, create the potential for conflicts of interest, exploitation, and inappropriate boundaries in social workers' relationships with clients. Social workers should explore and may participate in bartering only in very limited circumstances when it can be demonstrated that such arrangements are an accepted practice among professionals in the local community, considered to be essential for the provision of services, negotiated without coercion, and entered into at the client's initiative and with the client's informed consent. Social workers who accept goods or services from clients as payment for professional services assume the full burden of demonstrating that this arrangement will not be detrimental to the client or the professional relationship.

(c) Social workers should not solicit a private fee or other remuneration for providing services to clients who are entitled to such available services through the social workers' employer or agency.

1.14 Clients Who Lack Decision-Making Capacity   When social workers act on behalf of clients who lack the capacity to make informed decisions, social workers should take reasonable steps to safeguard the interests and rights of those clients.

1.15 Interruption of Services   Social workers should make reasonable efforts to ensure continuity of services in the event that services are interrupted by factors such as unavailability, relocation, illness, disability, or death.

1.16 Termination of Services   (a) Social workers should terminate services to clients and professional relationships with them when such services and relationships are no longer required or no longer serve the clients' needs or interests.

(b) Social workers should take reasonable steps to avoid abandoning clients who are still in need of services. Social workers should withdraw services precipitously only under unusual circumstances, giving careful consideration to all factors in the situation and taking care to minimize possible adverse effects. Social workers should assist in making appropriate arrangements for continuation of services when necessary.

(c) Social workers in fee-for-service settings may terminate services to clients who are not paying an overdue balance if the financial contractual arrangements have been made clear to the client, if the client does not pose an imminent danger to self or others, and if the clinical and other consequences of the current nonpayment have been addressed and discussed with the client.

(d) Social workers should not terminate services to pursue a social, financial, or sexual relationship with a client.

(e) Social workers who anticipate the termination or interruption of services to clients should notify clients promptly and seek the transfer, referral, or continuation of services in relation to the clients' needs and preferences.

(f) Social workers who are leaving an employment setting should inform clients of appropriate options for the continuation of services and of the benefits and risks of the options.

2. Social Workers' Ethical Responsibilities to Colleagues

2.01 Respect   (a) Social workers should treat colleagues with respect and should represent accurately and fairly the qualifications, views, and obligations of colleagues.

(b) Social workers should avoid unwarranted negative criticism of colleagues in communications with clients or with other professionals. Unwarranted negative criticism may include demeaning comments that refer to colleagues' level of competence or to individuals' attributes such as race, ethnicity, national origin, color, sex, sexual orientation, age, marital status, political belief, religion, and mental or physical disability.

(c) Social workers should cooperate with social work colleagues and with colleagues of other professions when such cooperation serves the well-being of clients.

2.02 Confidentiality   Social workers should respect confidential information shared by colleagues in the course of their professional relationships and transactions. Social workers should ensure that such colleagues understand social workers' obligation to respect confidentiality and any exceptions related to it.

2.03 Interdisciplinary Collaboration   (a) Social workers who are members of an interdisciplinary team should participate in and contribute to decisions that affect the well-being of clients by drawing on the perspectives, values, and experiences of the social work profession. Professional and ethical obligations of the interdisciplinary team as a whole and of its individual members should be clearly established.

(b) Social workers for whom a team decision raises ethical concerns should attempt to resolve the disagreement through appropriate channels. If the disagreement cannot be resolved, social workers should pursue other avenues to address their concerns consistent with client well-being.

2.04 Disputes Involving Colleagues   (a) Social workers should not take advantage of a dispute between a colleague and an employer to obtain a position or otherwise advance the social workers' own interests.

(b) Social workers should not exploit clients in disputes with colleagues or engage clients in any inappropriate discussion of conflicts between social workers and their colleagues.

2.05 Consultation   (a) Social workers should seek the advice and counsel of colleagues whenever such consultation is in the best interests of the clients.

(b) Social workers should keep themselves informed about colleagues' areas of expertise and competencies. Social workers should seek consultation only from colleagues who have demonstrated knowledge, expertise, and competence related to the subject of the consultation.

(c) When consulting with colleagues about clients, social workers should disclose the least amount of information necessary to achieve the purposes of the consultation.

2.06 Referral for Services (a) Social workers should refer clients to other professionals when the other professionals' specialized knowledge or expertise is needed to serve clients fully or when social workers believe that they are not being effective or making reasonable progress with clients and that additional service is required.

(b) Social workers who refer clients to other professionals should take appropriate steps to facilitate an orderly transfer of responsibility. Social workers who refer clients to other professionals should disclose, with clients' consent, all pertinent information to the new service providers.

(c) Social workers are prohibited from giving or receiving payment for a referral when no professional service is provided by the referring social worker.

2.07 Sexual Relationships (a) Social workers who function as supervisors or educators should not engage in sexual activities or contact with supervisees, students, trainees, or other colleagues over whom they exercise professional authority.

(b) Social workers should avoid engaging in sexual relationships with colleagues when there is potential for a conflict of interest. Social workers who become involved in, or anticipate becoming involved in, a sexual relationship with a colleague have a duty to transfer professional responsibilities, when necessary, to avoid a conflict of interest.

2.08 Sexual Harassment Social workers should not sexually harass supervisees, students, trainees, or colleagues. Sexual harassment includes sexual advances, sexual solicitation, requests for sexual favors, and other verbal or physical conduct of a sexual nature.

2.09 Impairment of Colleagues (a) Social workers who have direct knowledge of a social work colleague's impairment that is due to personal problems, psychosocial distress, substance abuse, or mental health difficulties and that interferes with practice effectiveness should consult with that colleague when feasible and assist the colleague in taking remedial action.

(b) Social workers who believe that a social work colleague's impairment interferes with practice effectiveness and that the colleague has not taken adequate steps to address the impairment should take action through appropriate channels established by employers, agencies, NASW, licensing and regulatory bodies, and other professional organizations.

2.10 Incompetence of Colleagues (a) Social workers who have direct knowledge of a social work colleague's incompetence should consult with that colleague when feasible and assist the colleague in taking remedial action.

(b) Social workers who believe that a social work colleague is incompetent and has not taken adequate steps to address the incompetence should take action through appropriate channels established by employers, agencies, NASW, licensing and regulatory bodies, and other professional organizations.

2.11 Unethical Conduct of Colleagues (a) Social workers should take adequate measures to discourage, prevent, expose, and correct the unethical conduct of colleagues.

(b) Social workers should be knowledgeable about established policies and procedures for handling concerns about colleagues' unethical behavior. Social workers should be familiar with national, state, and local procedures for handling ethics complaints. These include policies and procedures created by NASW, licensing and regulatory bodies, employers, agencies, and other professional organizations.

(c) Social workers who believe that a colleague has acted unethically should seek resolution by discussing their concerns with the colleague when feasible and when such discussion is likely to be productive.

(d) When necessary, social workers who believe that a colleague has acted unethically should take action through appropriate formal channels (such as contacting a state licensing board or regulatory body, an NASW committee on inquiry, or other professional ethics committees).

(e) Social workers should defend and assist colleagues who are unjustly charged with unethical conduct.

3. Social Workers' Ethical Responsibilities in Practice Settings

3.01 Supervision and Consultation (a) Social workers who provide supervision or consultation should have the necessary knowledge and skill to supervise or consult appropriately and should do so only within their areas of knowledge and competence.

(b) Social workers who provide supervision or consultation are responsible for setting clear, appropriate, and culturally sensitive boundaries.

(c) Social workers should not engage in any dual or multiple relationships with supervisees in which there is a risk of exploitation of or potential harm to the supervisee.

(d) Social workers who provide supervision should evaluate supervisees' performance in a manner that is fair and respectful.

3.02 Education and Training (a) Social workers who function as educators, field instructors for students, or trainers should provide instruction only within their areas of knowledge and competence and should provide instruction based on the most current information and knowledge available in the profession.

(b) Social workers who function as educators or field instructors for students should evaluate students' performance in a manner that is fair and respectful.

(c) Social workers who function as educators or field instructors for students should take reasonable steps to ensure that clients are routinely informed when services are being provided by students.

(d) Social workers who function as educators or field instructors for students should not engage in any dual or multiple relationships with students in which there is a risk of exploitation or potential harm to the student. Social work educators and field instructors are responsible for setting clear, appropriate, and culturally sensitive boundaries.

3.03 Performance Evaluation Social workers who have responsibility for evaluating the performance of others should fulfill such responsibility in a fair and considerate manner and on the basis of clearly stated criteria.

3.04 Client Records (a) Social workers should take reasonable steps to ensure that documentation in records is accurate and reflects the services provided.

(b) Social workers should include sufficient and timely documentation in records to facilitate the delivery of services and to ensure continuity of services provided to clients in the future.

(c) Social workers' documentation should protect clients' privacy to the extent that is possible and appropriate and should include only information that is directly relevant to the delivery of services.

(d) Social workers should store records following the termination of services to ensure reasonable future access. Records should be maintained for the number of years required by state statutes or relevant contracts.

3.05 Billing   Social workers should establish and maintain billing practices that accurately reflect the nature and extent of services provided and that identify who provided the service in the practice setting.

3.06 Client Transfer   (a) When an individual who is receiving services from another agency or colleague contacts a social worker for services, the social worker should carefully consider the client's needs before agreeing to provide services. To minimize possible confusion and conflict, social workers should discuss with potential clients the nature of the clients' current relationship with other service providers and the implications, including possible benefits or risks, of entering into a relationship with a new service provider.

(b) If a new client has been served by another agency or colleague, social workers should discuss with the client whether consultation with the previous service provider is in the client's best interest.

3.07 Administration   (a) Social work administrators should advocate within and outside their agencies for adequate resources to meet clients' needs.

(b) Social workers should advocate for resource allocation procedures that are open and fair. When not all clients' needs can be met, an allocation procedure should be developed that is nondiscriminatory and based on appropriate and consistently applied principles.

(c) Social workers who are administrators should take reasonable steps to ensure that adequate agency or organizational resources are available to provide appropriate staff supervision.

(d) Social work administrators should take reasonable steps to ensure that the working environment for which they are responsible is consistent with and encourages compliance with the *NASW Code of Ethics*. Social work administrators should take reasonable steps to eliminate any conditions in their organizations that violate, interfere with, or discourage compliance with the *Code*.

3.08 Continuing Education and Staff Development   Social work administrators and supervisors should take reasonable steps to provide or arrange for continuing education and staff development for all staff for whom they are responsible. Continuing education and staff development should address current knowledge and emerging developments related to social work practice and ethics.

3.09 Commitments to Employers   (a) Social workers generally should adhere to commitments made to employers and employing organizations.

(b) Social workers should work to improve employing agencies' policies and procedures and the efficiency and effectiveness of their services.

(c) Social workers should take reasonable steps to ensure that employers are aware of social workers' ethical obligations as set forth in the *NASW Code of Ethics* and of the implications of those obligations for social work practice.

(d) Social workers should not allow an employing organization's policies, procedures, regulations, or administrative orders to interfere with their ethical practice of social work. Social workers should take reasonable steps to ensure that their employing organizations' practices are consistent with the *NASW Code of Ethics*.

(e) Social workers should act to prevent and eliminate discrimination in the employing organization's work assignments and in its employment policies and practices.

(f) Social workers should accept employment or arrange student field placements only in organizations that exercise fair personnel practices.

(g) Social workers should be diligent stewards of the resources of their employing organizations, wisely conserving funds where appropriate and never misappropriating funds or using them for unintended purposes.

3.10 Labor–Management Disputes   (a) Social workers may engage in organized action, including the formation of and participation in labor unions, to improve services to clients and working conditions.

(b) The actions of social workers who are involved in labor–management disputes, job actions, or labor strikes should be guided by the profession's values, ethical principles, and ethical standards. Reasonable differences of opinion exist among social workers concerning their primary obligation as professionals during an actual or threatened labor strike or job action. Social workers should carefully examine relevant issues and their possible impact on clients before deciding on a course of action.

4. Social Workers' Ethical Responsibilities as Professionals

4.01 Competence   (a) Social workers should accept responsibility or employment only on the basis of existing competence or the intention to acquire the necessary competence.

(b) Social workers should strive to become and remain proficient in professional practice and the performance of professional functions. Social workers should critically examine and keep current with emerging knowledge relevant to social work. Social workers should routinely review the professional literature and participate in continuing education relevant to social work practice and social work ethics.

(c) Social workers should base practice on recognized knowledge, including empirically based knowledge, relevant to social work and social work ethics.

4.02 Discrimination   Social workers should not practice, condone, facilitate, or collaborate with any form of discrimination on the basis of race, ethnicity, national origin, color, sex, sexual orientation, age, marital status, political belief, religion, or mental or physical disability.

4.03 Private Conduct   Social workers should not permit their private conduct to interfere with their ability to fulfill their professional responsibilities.

4.04 Dishonesty, Fraud, and Deception   Social workers should not participate in, condone, or be associated with dishonesty, fraud, or deception.

4.05 Impairment   (a) Social workers should not allow their own personal problems, psychosocial distress, legal problems, substance abuse, or mental health difficulties to interfere with their professional judgment and performance or to jeopardize the best interests of people for whom they have a professional responsibility.

(b) Social workers whose personal problems, psychosocial distress, legal problems, substance abuse, or mental health difficulties interfere with their professional judgment and performance should immediately seek consultation and take appropriate remedial action by seeking professional help, making adjustment in workload, terminating practice, or taking any other steps necessary to protect clients and others.

4.06 Misrepresentation (a) Social workers should make clear distinctions between statements made and actions engaged in as a private individual and as a representative of the social work profession, a professional social work organization, or the social worker's employing agency.

(b) Social workers who speak on behalf of professional social work organizations should accurately represent the official and authorized positions of the organizations.

(c) Social workers should ensure that their representations to clients, agencies, and the public of professional qualifications, credentials, education, competence, affiliations, services provided, or results to be achieved are accurate. Social workers should claim only those relevant professional credentials they actually possess and take steps to correct any inaccuracies or misrepresentations of their credentials by others.

4.07 Solicitations (a) Social workers should not engage in uninvited solicitation of potential clients who, because of their circumstances, are vulnerable to undue influence, manipulation, or coercion.

(b) Social workers should not engage in solicitation of testimonial endorsements (including solicitation of consent to use a client's prior statement as a testimonial endorsement) from current clients or from other people who, because of their particular circumstances, are vulnerable to undue influence.

4.08 Acknowledging Credit (a) Social workers should take responsibility and credit, including authorship credit, only for work they have actually performed and to which they have contributed.

(b) Social workers should honestly acknowledge the work of and the contributions made by others.

5. Social Workers' Ethical Responsibilities to the Social Work Profession

5.01 Integrity of the Profession (a) Social workers should work toward the maintenance and promotion of high standards of practice.

(b) Social workers should uphold and advance the values, ethics, knowledge, and mission of the profession. Social workers should protect, enhance, and improve the integrity of the profession through appropriate study and research, active discussion, and responsible criticism of the profession.

(c) Social workers should contribute time and professional expertise to activities that promote respect for the value, integrity, and competence of the social work profession. These activities may include teaching, research, consultation, service, legislative testimony, presentations in the community, and participation in their professional organizations.

(d) Social workers should contribute to the knowledge base of social work and share with colleagues their knowledge related to practice, research, and ethics. Social workers should seek to contribute to the profession's literature and to share their knowledge at professional meetings and conferences.

(e) Social workers should act to prevent the unauthorized and unqualified practice of social work.

5.02 Evaluation and Research (a) Social workers should monitor and evaluate policies, the implementation of programs, and practice interventions.

(b) Social workers should promote and facilitate evaluation and research to contribute to the development of knowledge.

(c) Social workers should critically examine and keep current with emerging knowledge relevant to social work and fully use evaluation and research evidence in their professional practice.

(d) Social workers engaged in evaluation or research should carefully consider possible consequences and should follow guidelines developed for the protection of evaluation and research participants. Appropriate institutional review boards should be consulted.

(e) Social workers engaged in evaluation or research should obtain voluntary and written informed consent from participants, when appropriate, without any implied or actual deprivation or penalty for refusal to participate; without undue inducement to participate; and with due regard for participants' well-being, privacy, and dignity. Informed consent should include information about the nature, extent, and duration of the participation requested and disclosure of the risks and benefits of participation in the research.

(f) When evaluation or research participants are incapable of giving informed consent, social workers should provide an appropriate explanation to the participants, obtain the participants' assent to the extent they are able, and obtain written consent from an appropriate proxy.

(g) Social workers should never design or conduct evaluation or research that does not use consent procedures, such as certain forms of naturalistic observation and archival research, unless rigorous and responsible review of the research has found it to be justified because of its prospective scientific, educational, or applied value and unless equally effective alternative procedures that do not involve waiver of consent are not feasible.

(h) Social workers should inform participants of their right to withdraw from evaluation and research at any time without penalty.

(i) Social workers should take appropriate steps to ensure that participants in evaluation and research have access to appropriate supportive services.

(j) Social workers engaged in evaluation or research should protect participants from unwarranted physical or mental distress, harm, danger, or deprivation.

(k) Social workers engaged in the evaluation of services should discuss collected information only for professional purposes and only with people professionally concerned with the information.

(l) Social workers engaged in evaluation or research should ensure the anonymity or confidentiality of participants and of the data obtained from them. Social workers should inform participants of any limits of confidentiality, the measures that will be taken to ensure confidentiality, and when any records containing research data will be destroyed.

(m) Social workers who report evaluation and research results should protect participants' confidentiality by omitting identifying information unless proper consent has been obtained authorizing disclosure.

(n) Social workers should report evaluation and research findings accurately. They should not fabricate or falsify results and should take steps to correct any errors later found in published data using standard publication methods.

(o) Social workers engaged in evaluation or research should be alert to and avoid conflicts of interest an dual relationships with participants, should inform participants when a real or potential conflict of interest arises, and should take steps to resolve the issue in a manner that makes participants' interests primary.

(p) Social workers should educate themselves, their students, and their colleagues about responsible research practices.

6. Social Workers' Ethical Responsibilities to the Broader Society

6.01 Social Welfare Social workers should promote the general welfare of society, from local to global levels, and the development of people, their communities, and their environments. Social workers should advocate for living conditions conducive to the fulfillment of basic human needs and should promote social, economic, political, and cultural values and institutions that are compatible with the realization of social justice.

6.02 Public Participation   Social workers should facilitate informed participation by the public in shaping social policies and institutions.

6.03 Public Emergencies   Social workers should provide appropriate professional services in public emergencies to the greatest extent possible.

6.04 Social and Political Action   (a) Social workers should engage in social and political action that seeks to ensure that all people have equal access to the resources, employment, services, and opportunities they require to meet their basic human needs and to develop fully. Social workers should be aware of the impact of the political arena on practice and should advocate for changes in policy and legislation to improve social conditions in order to meet basic human needs and promote social justice.

(b) Social workers should act to expand choice and opportunity for all people, with special regard for vulnerable, disadvantaged, oppressed, and exploited people and groups.

(c) Social workers should promote conditions that encourage respect for cultural and social diversity within the United States and globally. Social workers should promote policies and practice that demonstrate respect for difference, support the expansion of cultural knowledge and resources, advocate for programs and institutions that demonstrate cultural competence, and promote policies that safeguard the rights of and confirm equity and social justice for all people.

(d) Social workers should act to prevent and eliminate domination of, exploitation of, and discrimination against any person, group, or class on the basis of race, ethnicity, national origin, color, sex, sexual orientation, age, marital status, political belief, religion, or mental or physical disability.

Chapter 2

Case Management: Definition and Responsibilities

INTRODUCTION

==Case management is one of the primary places in human service systems where the whole person is taken into account.== Unlike specific services, case management does not focus on just one problem, but rather on the many issues, strengths, and concerns a client brings.

For instance, an elderly person may be referred to Help Ministries for a voucher for fuel oil because it has been unusually cold and the elderly person has been unable to pay for the additional oil needed to warm his home adequately. In this case, Help Ministries is concerned with his fuel oil need and the warmth he will need to stay in his home during the winter. That is their only concern with regard to this client.

The case manager, on the other hand, is concerned with the client's need for fuel oil, with his desire to move into public housing for the elderly in the spring, with what resources he has among his children, with his recent slurred speech indicating a possible stroke, and with his need for Meals-on-Wheels. The case manager is aware that there is a neighbor who can look in on her client daily, that the client has ties to a church, and that he receives Social Security, but little other income. She knows he has a sense of humor, goes to Bingo once a month, and should be fitted for a cane.

==Case management is a process for assessing the client's total situation and addressing the needs and problems found in that assessment.== As a part of this process, the client's strengths and interests are used to improve the overall situation wherever possible. The primary purpose for case management is to improve the quality of life for the client. This might mean more comfortable or safer living arrangements, or it might require psychiatric care or medication for diabetes. Another major purpose of this activity is to prevent problems from growing worse and costing more to remedy in the future. In the situation of the elderly man just described, we find that the Meals-on-Wheels program will deliver a certain standard of good nutrition, preventing malnutrition and costly medical bills in the future. By getting the client a cane, we may be preventing falls that would shorten his life and cost much more in medical bills to repair his injuries. If we enlist the neighbor to look in on our client everyday, we have provided a link between the client and his neighborhood. In addition, the neighbor can alert us to small problems that require our attention.

WHY CASE MANAGEMENT?

Case management serves two purposes. First, it is a method for determining an individualized service plan for each client and monitoring that plan to be sure it is effective. Second, it is a process used to ensure that the money being spent for the client's services is being spent wisely and in the most efficient manner.

The money you oversee in client care may be public money, such as the money that comes from the state to a county to administer mental health services. It may be money that is provided by insurance companies for services to a policyholder. It may be money provided directly to an agency from either of these sources for the care of a client. The agency uses case managers to make certain the most effective use is made of the money. It is, therefore, the case manager who determines what is needed and how to prevent needs and problems from escalating. It is the case manager who, in collaboration with others, determines what services should be authorized with the existing money. It is the case manager who then follows the client and the client's services and treatment to keep the plan on track.

Case management is more than looking out for another entity's money. It is also the most efficient way to make certain a client receives the most individualized plan for service and treatment possible. To ensure that this will be done, case management responsibilities have been broken into four basic categories of service: assessment, planning, linking, and monitoring. We will look at those categories in the order in which they are usually accomplished when working with a person.

WHAT IS AN INDIVIDUALIZED PLAN?

After you have worked with clients to determine where the problems are and what areas need attention, you will also know about the supports and other resources the clients have in the community and among their family members and friends. You will know what they do well and what interests them most. Each client will be different.

As you go about designing the plan for a person, you will place in that plan elements that take advantage of the client's strengths and supports. In addition, you will address those problems most outstanding or immediate for that client. Because each client has a different set of strengths, life circumstances, and immediate problems and no two clients view their situations in exactly the same way, no two plans will be exactly alike. Each plan will be developed specifically for the individual client.

At one time, a small program for homeless women employed a part-time case manager for the children. Homeless women were given two years in apartments belonging to the program to work hard on getting an education, training, and a stable source of income. Many of them were distracted from this by concerns about their children. Still others had little time to think about what their children needed as they went about restructuring their lives.

The county mental health/mental retardation program gave the shelter a small stipend to hire a children's case manager. The shelter hired a young woman who had just graduated from college. This seemed like an ideal choice. She was energetic, related well to the children, and was genuinely concerned about each of them. In the next year, the program monitor from the county noticed two things. First, there seemed to be very little material on the children in any records. No individual plans could be found, and no assessments on each child appeared to have been done. Second, the children were all following much the same plan. All the girls attended gymnastics. All the boys were enrolled in Little League. On certain weekends all the children went to the zoo or to the circus.

After receiving repeated requests for individualized plans for each child and some guidance about how to create them, the case manager quit. She said, on departing, that she did not have time to sit and write up records, that the children had been "having fun," and that the county was unreasonable. The county became more involved in the hiring of the second case manager, and this person was well aware of the importance of individualized planning.

In the first six months, two children began to get orthodontic work done, one received a scholarship to a private school, four boys went to Little League, one took violin lessons, and a third joined the swim team at the YMCA. Most of the younger children went to the circus and to the zoo. Most of the older children went on a bus trip to Washington, DC, and half of them went to two symphony orchestra performances that winter. No child's plan was the same as that of another child. Each child's needs had been documented and addressed in some way.

In developing these plans, the case manager called all her contacts in the community. She asked two dentists to donate their time. She prevailed upon the symphony to give her the tickets for two performances. She went to a private school and talked to them about this particularly gifted child until a plan for financing the child's education was worked out. She found a violin teacher and asked for fifteen free lessons as a gift to the shelter. In churches and synagogues, she got people enthused about helping the children whose mothers were working so hard to put a stable life together for their families. She looked at scout troops, church youth groups, and organized sports for possible answers to the children's needs. In any number of cases, the plan simply involved the case manager helping an older child choose from among school activities and arranging transportation.

This is what is meant by individualized planning. When it is done well and done creatively, your clients can grow and thrive.

ASSESSMENT

The first case management task is assessment. Good case managers do comprehensive assessments of their clients. All the client's strengths, needs, and weaknesses are part of the assessment. In addition, potential problems and strengths are considered.

An initial assessment that is thorough and comprehensive accomplishes this task. Through a series of questions, the case manager assembles information that presents a profile of the client. In this profile, strengths and weaknesses emerge. During the assessment, the case manager and the client begin to prioritize the problems that need attention.

A good assessment is the foundation for the development of an individual plan for service or treatment. It delineates the provision of essential services and individualized treatment. We will discuss in some detail later in this book how to do initial assessments.

PLANNING

After the assessment, you will be expected to develop an initial plan for the client that is comprehensive and addresses all the issues raised in your assessment. This plan should show incremental steps toward improvement and expected outcomes. As a case manager, you cannot plan well for your client unless you are thoroughly aware of the services, social activities, and resources in your community.

Formal Agencies

You want to begin by knowing what formal agencies are out there to help with a particular issue. For instance, if your client has a mental health problem, you might refer him to an agency that specializes in mental health treatment. The staff at that agency is familiar with medications, diagnoses, and treatment alternatives for mental health problems. Another client may be elderly and in need of social activities to occupy her time. You would refer her to a senior citizen center where recreational activities are planned for each day. A third client may be a child with academic problems in school. You might refer him to the free tutoring at a local church.

Generic Resources

Good planning is not limited to formal agencies, however. You want to know what resources are there for common problems we all have. Not every problem a person with a developmental disability has needs to be treated by agencies set up exclusively for individuals with developmental disabilities. For example, a woman with mild mental retardation, grieving the death of her mother, was welcomed into a grief support group at the local church and given much support.

Informal Resources and Folk Support Systems

You also need to be aware of social activities your clients might enjoy that would keep them involved in their community. Perhaps one likes to work on models and could become a member of the model railroaders club. Perhaps another genuinely likes people and enjoys being with them. This person might do well as a member of the Jaycees.

Clients do better living in a community where they have healthy folk support systems. A folk support system refers to the kinds of supports most of us have in our communities—Lions Club, a church, volunteering with others as fire fighters. All of us need to feel a part of the place where we live; but for many people, the skills to interact with others and find useful activities are not fully developed. As a case manager, it is your responsibility to integrate your client into the community if this is a need. Find social clubs, churches, and groups that pursue similar interests and help your client make contact with those people. The more contacts your client has and the more useful activities your client engages in, the more support the community can give.

Case managers often fail to use these valuable informal resources for several reasons. They may feel that their client cannot handle being with ordinary people in ordinary settings. This is often based on the case manager's attitude about the client's disability and is often quite erroneous. Having clients in small numbers in social activities or organizations that give them an opportunity to practice strengths is an invaluable experience for everyone concerned. Another reason a case manager might be reluctant to place a client in a community social group might stem from the case manager's perception that people in such groups do not want to be bothered with people who have disabilities. In some cases, this assessment is correct, but in others it is quite the opposite. Many organizations are set up to provide service and perceive this as an opportunity to grow and serve the community.

Doing your homework pays off. You cannot rely on suppositions and speculations. Know what is available in your community and have places in mind that would serve your clients as the need arises. Meet people, talk to them about what you would like to have available for your clients. Gradually you will develop a list of people and places that welcome your clients and provide the specific experiences and support you are seeking. Your task is to have at your fingertips many resources you can use in developing a plan for your client and to be continually developing new ones in your community.

Continued Planning

Continued planning, as you follow the case, will take into account changes the client may face. An example will illustrate this kind of planning, which you may be called upon to provide. Mary Beth has a mild developmental disability and was assigned to you when she left a state-run institution for individuals with mental retardation. When you did the intake planning, you determined that she would do better initially in a sheltered living arrangement for about a year. Because the goal is for her to move to an apartment of her own at the end of the year, your planning should start well in advance of this move. This planning makes the transition easier for her and for you. There are no shocks and sudden surprises that might necessitate her need for hospitalization or a regression back to greater dependence on the agency.

You might begin by setting up services and activities that involve her in her community. Mary Beth told you when you first talked to her of her interest in singing. The people at the institution said she loved music and sang well, although she could not read music. At the time Mary Beth came out of the institution, you could not find a good place for her to use her musical interests, but you noted this as a strength and kept your eyes open for an appropriate link. Now you have found a choir director at a small church who is willing to have Mary Beth sing with her choir. The church has numerous activities, and there are members who see to it that Mary Beth is included. In this way, you begin to prepare her for a move to more independent living. You seek and find a place for her to live not too far from the church, and you work with interested members to ensure that Mary Beth will have their continued help with transportation and inclusion in church activities.

You may be concerned that Mary Beth has other social ties to her community. There is the Aurora Club, created by professionals just for people with mental illness and developmental disabilities. This club is a place to go and meet others, and the club takes trips, goes bowling, and goes out to dinner together. You could refer your client there; however, you might decide that Mary Beth's mild disability does not warrant her being limited to social activities only for individuals with mental retardation. Instead, you might develop a relationship with a local women's club, getting them to take your client as a member.

As Mary Beth makes an adjustment to being outside the institution, you look for job placement. You make a referral to Goodwill, where she is able to develop her social skills, and soon she is hired by a local Wal-Mart as a greeter.

By the time Mary Beth moves into an apartment of her own, she has gained new confidence and many friends who connect her to the community. Her success is due in large measure to both your wise initial plan and your modifications of the plan as Mary Beth grew more independent.

LINKING

Once the plan is drawn up, the case manager links (or transfers or refers) the client to the service or persons who will carry out the plan. Linking a client to a specific service requires care and skill on your part. You need to know the best service that will meet the individual issues and needs of your client.

Linking your client to a social service agency that provides a specific service—such as day treatment, drug rehabilitation, or groups for victims of violent crime—will require a written referral. You will state why you are making the referral, indicating the problem for which the referral is being made and the goal that you expect as a result of your client's contact with the agency.

The referral will also indicate the amount of time you estimate it will take for the agency to reach this goal. The time limit is very important. It keeps treatment from becoming endless and unstructured. With a goal and a set amount of time in which to attain that goal, both the agency and the client are more likely to make the most of their time together.

Sometimes the client can take advantage of services on his or her own. You might tell a client about the Aurora Club, for example, and the next week he may take a bus there and begin going to the club regularly, participating in activities and social events. At other times, you may have clients who are unable to take the first step and who will need you to accompany them or to arrange transportation for them.

If the agency is a formal social service agency, personnel at that agency will be able to support your client in his or her program, and implement the goals and work on the issues you and the client have identified as important. Some agencies with very fine programs or specialized services are small, which may require you to give more support to the client. For example, at New Start, a staff of three focuses on second-stage groups for victims of rape and domestic violence. Much of the work done there is done by volunteers. The success rate is excellent, and the clients report a high degree of satisfaction with the agency; but the staff is not equipped to handle other problems that might develop while your client is in group. If you refer a client to a group at New Start and your client has landlord problems between group sessions, the staff at New Start may not be aware of it in time to prevent an eviction. Even if they become aware of it, they will need to refer the matter to you to help the client resolve the matter because of the limited staff time available to clients. On the other hand, at Riverview, a day treatment program, there are nurses who are aware of medication problems, social workers who monitor progress toward goals, and staff who can work to prevent an eviction of a client, if that appears imminent.

On occasion you may find a service for your client at an agency that does not seem interested in serving her. Perhaps they are reluctant because your client has been ill recently or because the agency is not interested in her type of case. In such instances, linking becomes advocacy as you advocate for your client or on behalf of your client. In a situation like this, advocacy means you will attempt to seek the best services for your client and you will insist that your client be treated fairly and with respect.

MONITORING

After the plan is made and it is implemented (meaning the referrals indicated in your plan have been accomplished), it becomes your responsibility to monitor the services given to your client. When a formal agency is holding a planning conference or treatment conference about one of your clients, you should be invited to attend. You should also receive written reports about your clients' progress and about the services given to them. If you do not receive reports at specified intervals from the agency, you need to contact them yourself on a regular basis.

Talking with another agency about the service they are giving your client is done for two reasons:

1. To be certain that the treatment or service you authorized for this client is in fact the treatment or service that is being given
2. To keep track of your client's progress toward the goals you developed with him or her and to be aware of times when modifications and revisions in either the goals or the plan need to take place

Less formal groups or institutions that are part of your plan should get a call or visit from you occasionally to monitor how the plan is working. Suppose that the neighbor offers to take your client, Bill, to church with her family every Sunday. In August, the family goes away for a month and does not make arrangements with anyone else to take him to church. He begins to feel lonely, and one day he goes to another church closer to his apartment. There he is extremely friendly to everyone, which seems to bother the minister and several members of the church. They decide he is "inappropriate" and call crisis intervention, which gets tied up going to the church and sorting out what happened. All of this could have been avoided if you had been able to have regular contact with the family who took your client to church. In that case, you would have known of the vacation and could have requested that they find a substitute or could have found a substitute yourself.

SKILLS YOU NEED TO DO CASE MANAGEMENT

Here are some of the skills you need to possess in order to do case management:

1. Knowledge of *individual and family dynamics* (which you find in courses such as Human Development, Introduction to Psychology, Marriage and the Family, and Abnormal Psychology)
2. Knowledge of the relationship between and among social, psychological, physiological, and economic factors (as found in the *ecological model,* a theoretical basis for evaluating a person's situation and needs)
3. Knowledge of the *focus and policies of your agency*
4. Knowledge of *state and federal laws and regulations* that affect your agency's delivery of service
5. Knowledge of the vast array of *community services and resources* where you practice

SOME USEFUL GUIDELINES

Here are some useful guidelines for you to follow in practicing case management:

1. Plan ahead. Plan before there is a crisis. Develop a plan that will prevent crises based upon what you learned about your clients in the assessment and what you can foresee happening in their situations if the issues are not addressed. Alleviate crisis-provoking situations before the crisis occurs.
2. Be accountable to your client and to the community. Do what you say you will do. Do it *promptly.* And carefully *document* what you have done.
3. Be optimistic about your client. Expect improvement and some degree of independence, and that is what you are most likely to get. Reinforce success, and

never miss an opportunity to give positive feedback. Set up situations in which your client is likely to succeed rather than situations that are complex and tricky.

4. Let your clients decide what issues and problems in their lives take priority. Involve your clients in all phases of the planning. Get their opinions and feedback about services and about the plan.

5. Go where your clients are. Do not stay shut up in your office. Go out and see where your clients are interacting with others, and teach your clients new skills in the field where they will need them.

6. Promote independence. Show pride in the independence your clients demonstrate regardless of how small it is. Model independence, encourage independence, and teach as often as it takes the skills to maintain independence.

7. Develop a large number of resources and be able to find good ones for your clients. Know what formal agencies exist in your community and their focus. Look for and develop good folk support systems on which you can rely. As you move about the community, look for new resources you can add to your list.

SKILLS YOU NEED TO BE AN EFFECTIVE CASE MANAGER

Here are some of the skills you need to be an effective case manager:

1. The ability to *work effectively with people* to promote their growth
2. The ability to *work with people* of various professions, paraprofessionals, the public, and clients and their families
3. The ability to *identify what your client needs*
4. The ability to *keep accurate and well-organized records*
5. The ability to *allow the client to take leadership* in planning services
6. The ability to *develop creative resources* within your community to meet client needs

LEVELS OF CASE MANAGEMENT

In some agencies, there are several levels of case management. Clients receive a level of case management commensurate with their need and ability to function. In this text, we will look at three levels of case management: administrative, resource coordination, and intensive. In some agencies, these categories might go by other names or there might be more than three levels of case management. The following sections provide examples of how case management services might be organized.

Administrative Case Management

This level of case management is assigned to high-functioning individuals who need very little assistance navigating the system. On occasion they might need a prescription refilled, an emergency appointment, or new services, but for the most part they are capable of handling these details themselves. These clients are placed in a pool with other clients who require little service or follow-up beyond the original referral. For the most part, the clients function independently, using well the services to which they were referred. When something does come up for a client in this caseload, an available caseworker handles it. This means that a client on this caseload does not always see the same case manager.

Resource Coordination

This next level of case management is reserved for individuals who have some trouble handling the details of their treatment or plan. They usually need help and may have more involved or chronic difficulties that require more assistance. They do not, however, pose a risk to themselves or to others. In addition, with good support, they are unlikely to experience repeated hospitalizations or other crises. Here caseloads are larger, and clients are often in need of services and assistance on issues like housing, medication, or therapy; but generally the clients do well with the services offered and do not pose a risk to themselves or others. A person going through a particularly difficult time might be moved up to intensive case management and then return to resource coordination after the stressful circumstances have been addressed.

Intensive Case Management

Individuals receiving intensive case management require considerable supervision and assistance in order to remain in the community and in circumstances that do not exacerbate their problems. Generally, the caseloads of intensive case managers are smaller, allowing for more individual attention. Clients in the caseload would be those at high risk for repeated emergencies and hospitalization or at risk of deteriorating to the point that they pose a danger to themselves or others. Intensive case management is usually available twenty-four hours a day and requires intense involvement with the client to ensure that the person has a support network available and is not in high-risk situations, such as running out of medications or living in a housing situation likely to trigger stress and relapse.

SEPARATING CASE MANAGEMENT FROM THERAPY

Case management is not therapy. Often beginning case managers believe that they are to do therapy; that is, that they are to provide weekly talking sessions in which deep-seated conflicts and concerns are exposed and resolved. That is not the purpose of case management; and indeed, most case managers are not prepared to handle this type of work.

In the course of case management with a client, you may uncover deep-seated problems and issues. These become the basis of a piece of the plan developed to resolve these issues. The client is referred to a person or agency that can do that work expertly.

As the case manager, you may be the one your clients call when they are having a crisis and their therapist is unavailable. Good listening skills and helping the clients develop a way to handle things until they are seen in therapy is the case manager's role. It is not your role to intervene with a therapy session.

Finally, you will find plenty of other problems that do call for innovative interventions on your part. Learning to be independent, adopting useful and appropriate work habits, practicing good interpersonal skills, and behaving appropriately are all areas that you may address with your client in the course of case management. Although the client may be referred to a specific agency for exactly those skills and that information, you will want to support that intervention in your contacts with the person.

Many clients do not require therapy. Perhaps they have had considerable therapy in the past and are not able to benefit from it now or were never able to benefit from it.

Perhaps their interpersonal problems are more a result of the chemical imbalance they suffer than psychological dynamics. Maybe the client has a developmental disability and she only needs skills to attain as much independence as she is capable of handling. Some clients might have suffered a crisis such as rape or domestic violence and need a plan that focuses on protection and independence. Others may be out of fuel and need a plan that resolves the problem of a cold home with small children in it.

What case managers do is therapeutic in the sense that it benefits the client. Conducting clinical therapy, however—where a person comes in about long-standing emotional problems or pervasive affective disorders—takes years of study and training and should never be attempted by a person not specifically trained to conduct therapy.

EXERCISES IN CASE MANAGEMENT

Instructions: In each of the situations that follow, develop a tentative plan for the client. List the various services you believe each client needs initially. Include in your plan for each client both formal and informal services; and, where appropriate, use generic services and agencies. Suggest other services the client might use later, after the case is stabilized. Think about how you can involve others close to the client and how you will involve the client in planning.

1. You are called by the daughter of an elderly woman, who lives alone. The daughter lives in another city and is concerned because her mother does not drive and has seemed unhappy and listless on the phone. The daughter expresses concern that her mother seems lonely and is perhaps depressed. The daughter does not know her mother's neighbors and calls you instead at the Office of Aging. She has told her mother she is going to call your agency for help, and the mother had no objection to that. *Home visit of case management*

2. A man with a developmental disability lives alone with his widowed mother. She has fallen and broken her hip and will be at the rehabilitation hospital for about six weeks. He cannot stay alone. He has a job at Goodwill Industries. County transportation takes him there every morning at 8:30 and brings him home at 5:00 P.M.

3. A woman and her two children are waiting to receive their welfare check. They came to your state from another to escape an abusive husband and father. The

woman is frail and appears sick. They have no place to go and have not eaten in several days. The children smell as if they need a bath and are listless.

4. A man has been referred by his family physician for help. The man seems extremely inebriated. His wife brings him in and says she is worried that he may go into delirium tremens if he withdraws from alcohol too quickly. His family physician did not see him but sent the couple straight to your office.

5. A father brings in his 14-year-old daughter, who is running the streets, refusing to listen, and failing in school. He is at his wits' end, saying he must work and is not there when the girl returns from school. Her mother died four years ago, and the trouble started when the daughter was about 12. The father feels that he and his daughter have a difficult time communicating with one another.

6. A police officer asks you to come to the home of an older man he has been concerned about for several weeks now. The man is delighted to see you and tells you that he is having pains in his legs and is unable to walk. During your visit, he asks you to get things for him that are nearby, but obviously it is too painful for him to get up. He says he does not go to the kitchen often to prepare meals, but the police officer has stopped by several times with sandwiches, and Mrs. Jones from up the

street, an old friend of the man's late wife, has brought a casserole on occasion. He is adamant that he wants to stay in his home as long as he can.

Meal-on-wheel
Home Health Care
Doctor evaluation
Rehab
Family Support
Other Support system
Insurance
Financial Status

Transportation System
Follow-up Care
Church Support System

7. A woman comes in complaining of depression. She says it started when her husband left with a younger woman and she has not been "right since." She reports having difficulty falling asleep and complains of no appetite. She says she has missed over three weeks of work since he left last month. There are no children, but she tells you she has neglected the dog and cannot remember if she fed him last night or not. She appears listless and very sad, weeping off and on during the interview.

8. A mother brings her 12-year-old son in because they are "not getting along." She reports that he does not listen and comes and goes as he pleases. His homework has fallen off, and his grades have slipped, but he is still doing well in math and likes his math teacher. The boy's father was killed in a railroad accident two years ago. The mother tells you that the boy and his father enjoyed a close and warm relationship and that she has felt her influence on him slipping away since the accident.

9. A woman, recently placed in the community after three years in a state mental hospital, is having trouble adjusting to the living arrangement made for her by the hospital. She is not going out and does not participate in any activities. She is friendly

when you talk to her and seems glad to have your company, but she does not seem to know how to take care of the details of everyday living. She has a roommate who is more competent and independent. The two get along well.

10. A woman with two small children is referred to you because she recently lost her apartment. She has a meager income from a part-time job as a clerk in a convenience store and was unable to pay the rent and take care of other bills. She seems unaware that she might be eligible for financial assistance. She is not sure where the children's father is at the moment. All her belongings are packed in five bulging garbage bags. She and her children seem malnourished and thin.

Chapter 3

Applying the Ecological Model: A Theoretical Foundation for Human Services

INTRODUCTION

In working with other people, human service professionals apply the ecological model to develop a broad understanding of each individual client who comes before them. This model, sometimes referred to as person-in-situation or person-in-environment model, looks at the individual client in context.

You are well aware of how distorted communication can appear when a quote is given out of context. It is possible to skew impressions and deliberately create misunderstandings by quoting only a portion of what someone has said and not the entire conversation. For instance, when Matt was questioned in class about his homework assignment by his teacher, he said he was not sure if he could help people with a certain disability. He went on to explain that at one time it was thought that he had that disability and he had worked very hard to prove that he was not disabled and to overcome people's initial impressions of him. Now he found being with people who suffered from this disability uncomfortable. He also explained that he expected his course of study to help him overcome that problem and he was very aware of his reactions and the fact that they might be inappropriate.

Later Anne, who was in class that day, confided to Aisha and Alice that she did not feel Matt should be allowed to continue with his studies. Surprised, Aisha asked Anne why she felt that way. "Oh, because he said in class the other day that he feels uncomfortable around certain disabled people. I mean, if you can't work with disabled people you need to find something else to do." Aisha and Alice quickly agreed.

Anne's description of what Matt said was distorted because Matt's comments were repeated outside the context in which they were said. That kind of distortion takes place when we look at individuals out of context. In your work, every person you will see functions in a context, an environment. You cannot adequately understand that person without also being able to understand the context in which that person functions and interacts.

It is very tempting to overlook context. Many of us fall into the trap of thinking that *A* causes *B*. Juan lost his job because he is irresponsible. If we eliminate *A*, *B* will cease to be a problem. If we make Juan more responsible, he will not lose any more jobs. Or Jill's husband left because she is demanding. If we teach Jill better interpersonal skills, she will have better relationships. While helping Juan to become more job-ready and helping Jill to communicate better may very well be a positive part of your plan for them, this kind of understanding of their problems and assigning of solutions largely ignores the context in which these problems arose. It also makes it much easier to see the individual as responsible or to blame the individual for the problems he or she has brought to your attention. When we blame others, we nearly always feel less empathy with their difficulties and less inclined to be truly useful in the human service sense.

SEEKING A BALANCED VIEW OF THE CLIENT

All individuals constantly interact with any number of systems in their environments. All individuals bring to those interactions unique characteristics. Unless the human service worker has a balanced view of both the client and the client's context, important information and constructive opportunities are lost.

In the case of Ralph and Edwardo, we can see how important this is. Ralph had been in prison for some youthful gang activity. While he was there, he took advantage of every opportunity to change. He went to church regularly, developed a personal relationship with a minister who came to the prison often, and obtained his high school diploma. Ralph was a warm, humorous person who attracted many friends. His outgoing personality attracted people to him who ultimately encouraged him and gave him support. During his time in prison, his mother wrote to him often, pleading with him to change his ways. Ralph felt bad about the trouble he had caused his mother, particularly in view of the fact that she had raised him after his father left home, and he saw her letters as a reason to do better. When he left prison, he enrolled in college courses and attached himself to the church, where he was warmly welcomed.

Edwardo was in the same prison for youthful gang activities. He was quiet and retiring and did not attract the attention and support that Ralph had secured for himself. Edwardo attempted to get his high school diploma while in prison, but had trouble asking for help when he needed it and eventually gave the project up in frustration. Preferring not to join groups, he did not go to church or any other group activity that promoted independence and responsibility. Because Edwardo spoke so little and rarely smiled, he was often misunderstood as being hostile. In fact, he felt shy and awkward around other people. Edwardo's mother wrote him regularly, and she too pleaded with him to do better and "turn his life around"; but Edwardo tended to see these letters as nagging and to blame his mother for the fact that his father left when he was very young. He rarely answered her mail. When Edwardo left prison, he moved back with his old friends and resumed his former criminal activities.

This illustration demonstrates how individual characteristics play a role in the outcome for the client. Part of developing a balanced understanding of the client is being able to see what the client brings to the situation and how that interacts with the larger context of his or her life. Ralph brought a personality that attracted others to assist him. He brought a good relationship with his mother and a motivation to do things more constructively. Edwardo brought a more retiring personality, one that was less attractive to others and often misunderstood. Edwardo's interpersonal skills were not as developed as Ralph's. The individual characteristics of Edwardo and Ralph affected the outcome of their prison time.

Now we will look at Edwardo and Ralph differently. For our purposes, let us suppose that Ralph and Edwardo are both warm, humorous people. Both make friends easily and enjoy the company of other people. Each of them is sent to prison for youthful gang activities, but the context is different. Edwardo goes to a prison upstate. It was recently built and the focus is on rehabilitation. There Ed is provided with high school and college classes as well as religious and self-improvement activities. He is able to take advantage of many different programs to further his goals. A supportive counselor works with him to come up with a good set of goals and helps Ed implement these. They meet on a weekly basis. The location of the prison has another advantage. Edwardo is now closer to his father, who lives only a few miles from the prison. His father begins to visit, offering his support and a place for Ed to live when his sentence is completed. Edwardo leaves the prison on a solid footing and continues his work toward a college degree.

Ralph, on the other hand, is sent to an ordinary prison where the counseling staff is overwhelmed. His counselor sees Ralph's potential, but has trouble getting Ralph into high school courses because they are crowded. During the time Ralph is at the prison, the education staff experiences a number of turnovers and layoffs. Ralph never can get into the program and stick to it. He rarely sees his counselor because of the number of inmates with whom the counselor must work. No family member comes to visit Ralph, partly because he has been sent so far from them and partly because they blame him for his incarceration and have lost interest in him. Ralph's mother, sick with severe chronic asthma, rarely writes. There are church services, which Ralph attends regularly, but the prison does not allow inmates to meet the pastors before or after services because of a strict schedule, and the pastors who have formed relationships with inmates come irregularly at other times. When Ralph leaves the prison, he has not completed his high school diploma. He moves near some people he knew in prison, and soon he takes up the criminal activities in which he participated before his incarceration.

Here it is the context that is different. Edwardo finds himself in a supportive context: a counselor who focuses on his goals and sees that these are implemented, plenty of self-improvement opportunities, a warm relationship with his father, a prison committed to education. Ralph, however, finds himself confronted with indifference, lack of supportive programs and activities, an overwhelmed counselor, and a family too distant to give encouragement.

The interaction never ceases. The individual makes choices, but the environment has prompted those choices. The individual responds to the outcome of those choices, and the environment reacts or adapts to that response. This interaction begins at birth. A fussy baby with calm, patient parents will start life differently from a fussy baby with overworked, anxious parents. An infant with severe disabilities will receive a good start with a large, loving family who devote their time and energy to getting her the best medical and rehabilitative care. An infant with severe disabilities may arrive in another family where everyone tries their best to give the infant a good start; nevertheless, the disabilities prove overwhelming to the caretakers, there is no cure, and family members find that any semblance of a normal home life or time with other siblings is severely curtailed. The first baby is raised at home, while the second one is placed in a good institution.

In human services, the trained eye will look for and see a balanced view of the client and his or her context when assessing the individual's needs. Just as important will be the human service worker's understanding of how person and context interact to produce certain outcomes for the client.

THE THREE LEVELS

Client and context have been defined as having three levels:

1. *Micro level*, where the focus is on the client's personality, motivation, affect, and other personal attributes
2. *Meso level*, where the focus is on the context immediately surrounding the client (family, church group, close friends, and work group)
3. *Macro level*, where the focus is on the larger society's characteristics and the way the client experiences these or the way these are brought to bear on the client's situation (institutions and organizations like the political system, social stratification, the educational system, and the economy)

Human service workers are expected to be aware of all three levels when assessing a client's situation and to be able to intervene on all three levels where such intervention is appropriate.

LOOKING AT WHAT THE PERSON BRINGS

When you do an assessment to open a case or do follow-up planning, there are a number of individual characteristics that will impinge on the problems and the eventual outcome. These characteristics, which we place on the micro level, can be divided into two broad categories:

Biological Characteristics	Psychological Characteristics
Neurological development	Early shaping experiences
Reflexes	Perception
Genetic makeup	Personality
Degenerative processes	Affect
Illness (chronic, terminal, or temporary)	Cognition
Physical health	Nurturance
Nutrition	Life transitions/position in the life cycle
	Motivation

This general outline gives you a starting point for understanding what your client has brought to the situation. Clients have had different early life experiences, are composed of different genetic configurations, and possess different personalities and perceptions. Each of these differences interacts with the external circumstances of the client's situation to promote self-fulfillment and well-being, to block those goals, or, quite possibly, to have no effect at all.

Although we will look at this topic again in the chapter on assessment, this discussion puts forth the framework you will follow in assessing the individual's contribution to the situation.

LOOKING AT WHAT THE CONTEXT BRINGS

Your client will function in a context that is personal to him or her and in a larger context, which is the larger society. These contexts can be divided into two broad categories:

Personal Context (sometimes referred to as the mezzo level)	Social Context (sometimes referred to as the macro level)
Family	The larger culture of the society
Work group	The larger organization of the church or workplace
Social groups	The larger community
Family culture	Government
Family values	Economy
Family structure	Social class
Religious group	Prejudice and discrimination
Social stratification	
Role status, conflict, and strain	
Political system	

It is important to obtain information about the context in which clients grew up and in which they are now functioning. The reason for the client's problems may lie in the context rather than with the client. When you learn about the context, you learn more about what motivates your clients, what environmental cues they receive to behave or make decisions the way they do, and what early circumstances shaped their way of responding to their community and their situation. Clients come from different social contexts. Clients grew up in different households with different parents and levels of nutrition and encouragement. Clients have different ways of looking at things and explaining them. The economy may have favored their work or begun to dispense with it. The political system may have awarded your client's subgroup power or disenfranchised that group in some way. Your client may have experienced prejudice, an indifferent medical system, or a poor educational system. On the other hand, this person may have grown up in a wealthy suburb, attended private schools, and received the best medical care money could purchase.

DEVELOPMENTAL TRANSITIONS

The ecological model is also concerned with normal life changes, often referred to as transitions. These changes are called transitions because they are events that often move a person from one phase of life to another. These events require a person to make some adjustments or to adapt in some way to new circumstances. Many of these events are simply part of the normal development that all people experience from birth to death. These transitions are often expected, and for some transitions there is preparation.

Some people who go through a transition do not cope with the changes as well as others do. Perhaps there is more going on in their lives than they feel they can handle. Perhaps the event will change the person's life dramatically in ways that are viewed as negative. Many of the people we see are going through or have recently experienced one or more of these transitions. Let us look at a list of such transitions:

Starting kindergarten or first grade	Starting a new job
Going to high school	Getting married
Going out on the first date	Buying a first home
Leaving home for the first time	Experiencing ill health
Losing one's job	Losing a spouse through death
Experiencing a disaster	Divorce

A large mortgage or other debt	Losing some physical capacity
Considerable financial losses	Considerable financial gains
Children leave home	Children marry
Birth of grandchild	Death of a child

And there are many more.

Although every single one of those events probably will not happen to one person, we can expect that most people in the course of a long life will experience many of the transitions on the list. It is important to know about the common life stages a person passes through and to recognize where your clients are in their life stages and transitions. Often transition problems are common to so many people that there are self-help and support groups that do not entail treatment by a mental health professional, but rather support from others who have been through the same transition.

Sometimes, however, a person may find the changes overwhelming, completely negative, or intolerable. These people may need professional help to handle these changes and adjustments.

DEVELOPING THE INTERVENTIONS

You have looked carefully at your clients' issues and problems. You have come to understand the ways in which your clients have responded to their context and the way the context has contributed to your clients' motivations and decisions. As a case manager, your role is to design a plan with each client that will address the areas of need. Human service workers can and do intervene on several levels.

Many clients of the social welfare system appear to be individuals who have encountered inadequate support in their context. One or more of the institutions that we believe should support the individual in our society has failed these people or has been unable to supply what was necessary to avoid problems. Institutions such as education, medicine, the economy, politics, and the family may have let this person down in some way.

When this occurs, our society looks to the social welfare system to supply what is needed, to address the unfortunate gaps in a person's life, and to apply interventions that will prevent a worsening of the problems. Your task as case manager is to look at the client and the client's context, to gather the facts about each of these, and to understand how context and person interact to the detriment or the well-being of the client.

With this information, a plan is developed for the person that addresses maladaptive interactions between the individual and the environment and that notes those parts of the environment that are positive and useful. The interventions you design or choose would be two-pronged: personal interventions that strengthen the person to handle the environment, and environmental interventions that change the context to accommodate the person. Here are some examples of the types of interventions that can be incorporated into individual plans:

Personal Interventions	**Environmental Interventions**
1. AA for substance abuse	Family education to support sobriety
2. Parent skills training for the parents of a child they abused	Temporary removal from home to foster care
Group therapy for abused child	Foster home parents given information on creating supportive foster home environment

3. Job training for the person with a developmental disability

   Work with the employer to provide a supportive work environment

4. Interpersonal skills training for the adolescent in minor trouble

   Use his interests to develop more constructive after school activities
   Bring father into the picture in a positive way
   Family therapy with the mother

5. Medication for the person with schizophrenia
   Regular appointments with the psychiatrist

   Place in a supportive living environment in the community
   Develop constructive connections with the local church
   Two family sessions to reinvolve the family in the person's life

These are fairly routine interventions in a person's problems to ameliorate a negative or destructive situation or to enhance the person's self-fulfillment. The point is, however, that a service plan formed without the understanding that interventions take place in two distinct areas could be quite hapless and without focus. When doing a service plan, the human service worker makes certain that both areas have been addressed with appropriate interventions wherever possible and that the interventions are documented in the record.

LARGER INTERVENTIONS

It is assumed that the human service worker is not limited to just helping individuals. Your work as a human service professional places you in a unique situation to be able to speak to the problems affecting large numbers of people. This happens in two ways.

In the course of your work, you will encounter many who have been damaged by abuse or discrimination. You will see groups of people harmed by poor school systems, a lack of medical care, or scant supervision. Ethically we have an obligation to speak to the needs of those with less advantage in our society. Having seen the damage firsthand, we are better able to speak to conditions that need to be remedied in our larger society and to keep statistics and information on the extent of the problem for use in persuading lawmakers and others in power to take action.

You will also see areas of service that have been neglected or that require development. Perhaps there is a need for more supported living arrangements for those with mental illness in the community. It may be that mothers returning to work from welfare lack the means to dress appropriately for the job, need day care for their young children, or require transportation to get to job interviews. Who better than the human service professional to bring to the attention of those who develop programs those areas of service that are lacking in your immediate community? You and others can bring to light the unique needs of your community and help to develop much-needed services.

Assignment

Talk to a human service professional in your community to learn what that person sees as an unmet need in the community. Ask the professional what one service he or she would like to see developed. What specific need would that service meet? What client population would that project address? How could the clients who need this service be mobilized to work in their own behalf?

LOOKING AT FLORENCE'S PROBLEM ON THREE LEVELS

Instructions: Look at Florence's problem as she presented it to the case manager. Decide which parts of her problem are on the micro level, which parts are on the mezzo level, and which parts are on the macro level.

Florence came in to see a case manager in an agency that addresses child abuse and neglect. Recently her daughter, Crystal, was removed from the home because of complaints by neighbors that she was abusing the child. An investigation of the situation by child care workers indicated the abuse was severe. The discipline she was administering was discipline she had experienced and witnessed as a child from her own parents and her aunts and uncles who lived on farms near her family. Florence related that she was the oldest daughter, third in line of nine children, of a farm family of twelve people. Her parents worked hard from sunup until long after dark. Much of the housework was done by Florence and her aunt, who lived with them. Her mother was ill, often in her room in bed. Florence does not know what the illness was, but does not recall her mother ever seeing a doctor. She tells the case manager that she knows her mother and her aunt did not like her.

At 18, Florence ran away with Dave, who did mechanical work on cars. "He was my first and only boyfriend," she explains, weeping. Florence and Dave never married and they had one child, Crystal. Last April, Dave died in a car accident on the interstate. Florence cries as she describes that night and the way the police came to her trailer and how kind they were to her. She describes how alone she felt and has felt ever since.

Florence receives welfare. She completed eighth grade before her father "yanked me out of school to do housework. Said it was no place for a girl. A girl didn't need no schooling." Florence had enjoyed school, mostly for the companionship of other girls. "I'm shy of people, you know. But at school I had friends." Florence remembers school as hard, and she had trouble with subjects like math and science. "Mostly I sat there and worried about what would happen when I got home from school. It was always something, Mom was worse, I was in trouble, there was some big push to get in a harvest. I was glad when I quit."

Leaving with Dave had alienated Florence from her family. "Dave used to say, 'They're just mad 'cause they can't use you no more.'" For this reason, Florence has not seen her family since Dave's funeral, and they have made no attempt to get in touch with her even though they are only a few miles apart. The welfare agency reports that their workers have rarely seen Florence and have not as yet offered her any services for going to work, although she is on a list of single mothers they would like to make job-ready. Child welfare tells you that they cannot return Crystal until Florence has had intensive parent training and supervised visits with her child. They also tell you that they found her home worn, but immaculate.

Florence confides that she is terrified of going to work, that she feels useless, and that she probably has little to offer on a "real job." She also appears to be depressed, crying at intervals and hanging her head. Socially she is isolated both because of Dave's death and because her neighbors are fed up with her child care practices. "The neighbors don't like me either," she says with resignation. The child care agency is asking for parent training, but it is unclear who will offer that in this rural area.

What part of Florence's problem is a micro level problem?

What part of Florence's problem is a mezzo level problem?

What part of Florence's problem is a macro level problem?

DESIGNING THREE LEVELS OF INTERVENTION

Instructions: Look at the four cases below and decide how you would intervene on three levels: the personal, the contextual, and the larger environment.

1. Maria is paralyzed from the waist down following an accident three summers ago in a swimming pool. She is hoping to complete her degree in accounting, but is complaining of depression and an inability to focus on school. When you see her, she looks anxious and tired. Her affect is flat, and she tells you nothing interests her. At the local college, she has had trouble finding appropriate parking and misses many days of class when the weather is bad because of the parking situation. One of the professors she must work with rather closely has made remarks about the difficulty of "other people getting around that wheelchair." She believes her boyfriend, who was with her the night the accident happened, has remained with her simply out of pity. When they fight about other things, she throws this up to him, although he vehemently denies it and tells Maria this is a hurtful accusation.

    Interventions on the micro level:

    Interventions on the mezzo level:

    Interventions on the macro level:

2. Mr. Groff is 93 and living alone in his home. He only stopped driving last year. He would like to get out more, perhaps go to the senior citizen's center. In addition, he would like to go to the Lions Club and to participate in a foreign policy club he belonged to for years. He tells you sadly that the members of the foreign policy club always seemed amazed at the reading he had done and the sound opinions he expressed, "as though I should be senile!" Since he stopped driving, he has lost contact with them. Right now he sees no reason to go to a nursing home and feels that if he had transportation, he could continue to buy his groceries, prepare his meals,

and care for himself generally. He tells you, however, that he would like to find a way to be less lonely.

Interventions on the micro level:

Interventions on the mezzo level:

Interventions on the macro level:

3. Margie is in a sheltered workshop for people with developmental disabilities. She does well at work and has many friends. She lives alone with her mother, and her mother is not happy with the new level of independence Margie is developing. She often goes out with others from work and the supervisors for dinner on Friday night. She has joined a social group for individuals with disabilities much like hers, and they go bowling and to the movies. Margie has, since she went to the sheltered workshop, learned how to use the phone to make appointments with her doctor and dentist and how to ride the bus to and from both work and the social club, and she has been shopping to buy her own clothes twice with her case manager. When Margie's mother complains about all this, she tends to blame Margie for leaving her alone at night. "Since your father died, you're all I have," she tells Margie. Margie's response to this is to cry and stay in her room. Sometimes she has missed work, hoping to make her absences up to her mother.

Interventions on the micro level:

Interventions on the mezzo level:

Interventions on the macro level:

4. Chris is a single father who is trying to work and raise three small children. His wife was killed two years ago in a traffic accident. After the initial shock and outpouring of support from friends and neighbors, Chris found himself alone with all the responsibilities and very unsure of himself. He would like to meet other men who have the same problems, but cannot find any groups, even though he has been told about several men who are in the same situation. He tells you he is not sure what the best method is for disciplining his children, whom he describes as "good kids." Sometimes he feels he is too lenient with them, and at other times he is afraid he is unnecessarily strict with them. A local women's health center has groups for bereaved single parents, but Chris believes those would not be open to him. "It would be all women, wouldn't it?" he asks. In addition, he is having a hard time at work balancing the responsibilities there with parenting responsibilities at home. "Of course, I want to do a good job and get the promotions so I can support these kids through college, but I need to be home in the evening, or someone does, and I don't think that is always well received at work."

Interventions on the micro level:

Interventions on the mezzo level:

Interventions on the macro level:

Section 2

Useful Clarifications and Attitudes

 # Chapter 4
Cultural Competence

INTRODUCTION

Seeing each of our clients as unique individuals is the only way to accurately perceive them and to be constructive in the way we serve them. One key element of individuality is culture or subculture.

Most of our attitudes and perceptions are the result of our interactions with others throughout our lives. In time, these interactions come to seem natural. As professionals, we need to become aware of our personal ways of thinking about others and their situations. Is our thinking useful? Will it promote the well-being, self-esteem, and independence of our clients? Because we are a culturally diverse society, it becomes important for professionals in human services to respect differences and to seek to understand these differences whenever possible.

Culture and Communication

Each of us brings to any situation perceptions and attitudes that are influenced by our own culture. Our own ethnic group, family values, outstanding experiences, and cultural traditions all influence both the way we communicate to other people and what we believe other people mean when they communicate with us. Often we are unaware of the extent to which these factors color our interactions with other people.

In addition, we do not usually take the time to understand that others may come from a culture that may differ from our own significantly. We may judge others' actions by the standard prevalent in our own culture. We may expect certain behavior we believe is appropriate and become annoyed when we do not see that behavior. We may misunderstand the communication of others, leading to lost rapport and opportunities. This is dangerous when we have accepted the professional responsibility for giving assistance to other people.

Your Ethical Responsibility

Ethically, you have a responsibility to take the time and make the effort to become familiar with cultures that differ from your own with which you have extensive contact. It is not ethical to simply assume you know all there is to know about a group because you see members of that group on a daily basis. Instead, you need to ask questions, take seminars, and gather information that will enhance your understanding of that group or culture.

When You Are Not Sure

It is not possible, on the other hand, to study and become familiar with all the different cultures you might encounter in the course of your professional lifetime. In your work, it is quite likely that you may encounter someone from another group whose culture is unfamiliar to you and whom you will see only briefly. What you need is a method you can use that will allow you to participate in those encounters and interactions competently.

WHERE ARE THE DIFFERENCES?

Differences among people occur on a number of sociological levels. These differences can be overcome and understood, or they can become obstacles to good communication and understanding.

Cultures

Generally, cultures coincide with national or political boundaries. People living in one country have a culture that differs from the culture of those living just across the border. When we refer to culture, we are really talking about the culture assumed by an entire society.

This means that we in the United States have in common with one another a basic knowledge. We learned this knowledge through the socialization process—from our schools, parents, religions, and even television and magazines. Although each individual may see the culture just a bit differently and no one knows everything there is to know about it, people share enough in common to be able to relate to and cooperate with one another.

By the time we are young adults, the culture we carry with us in our heads is largely unconscious. Our culture influences how we communicate with other people, and it influences the way we determine what the other person means. In other words, what we say is affected by our culture, and, in turn, our interpretation of what another person is saying to us is colored by our culture. Because this process has become automatic for us, we are not aware of the significant influence our culture has on our interactions with others. Furthermore, if most of the time we are communicating with others from our own culture, we will assume that all people mean what we mean and see things as we see them. With this way of thinking firmly in place, there is a tendency to assume that our own culture is the better or correct way to be at any given time.

Subcultures

Within any given society, there are groups of individuals who, for the most part, follow the culture of their society, but hold in common with each other somewhat different cultural ideas. This may be a religious group that holds ideas that are somewhat different from mainstream thinking about patriotism and serving in the military. It might be an ethnic group whose subculture is shaped by the discrimination experienced in each generation.

Subcultures usually are not completely out of step with the larger society's culture. There is, however, something about the subgroup culture that sets its members apart. It might be values or traditions or beliefs. It might be lifestyle.

Race and Ethnic Group

Important in understanding subcultures is the understanding of the terms *race* and *ethnic group*. According to Gudykunst and Kim (1997)*, *race* refers to a "group of people who are biologically similar," and *ethnic group* refers to "a group of people who share a common cultural heritage usually based on a common national origin or language" (p. 20).

Often racial groups have distinguishing physical characteristics, while ethnic groups may be distinguished by their language, religion, or some other aspect of their culture. It is important to keep in mind that race alone is not a factor influencing communication. On the other hand, ethnicity with its culture can have considerable influence on communication.

If the racial or ethnic group of our clients indicates a subculture that is unfamiliar to us, the potential for misunderstanding is increased.

How We Develop a We-Versus-Them Attitude

During the process of socialization, we learn that some groups are acceptable and others are unacceptable. The acceptable groups are seen as in-groups by us. We are more comfortable with in-groups. We see the members as similar to us. We expect members of in-groups to hold beliefs and values very much like our own. We expect them to act as we would and to think as we do. When we talk about in-groups, we generally do so favorably, holding them in positive regard. We are better at predicting how members of in-groups will respond or behave.

Out-groups are those groups with whom we feel uncomfortable, groups with whom we have less inclination to interact on a regular basis. Generally we do not hold members of out-groups in a particularly favorable light. We may be suspicious of the motives of an out-group because we do not fully understand its culture. Using our own culture as the standard, we may find the out-group culture inferior. Members of the out-group may appear unpredictable, unreliable, or devious to us.

STRANGERS

When people do not act or think the way we believe they should, they seem strange to us. Many people you will encounter in the course of your work will seem like strangers to you. Look at Julio. Julio is a stranger to nearly everyone he sees on a daily basis. He, his little brother, and his mother live in a small city with others from Puerto Rico. When he is with his small group of friends and relatives, he is not perceived as strange. When, on the other hand, he attempts to interact with the larger American culture, many see him as a stranger. His language, his behavior, and, in some cases, his attitudes appear strange to members of the larger culture. He is tolerated, and even given menial work, but he feels set apart. People he must meet and work with every day view him as a stranger because they know little about Julio's culture. They see him every day, yet he is in no way considered part of their group.

Julio went to the case management unit to seek services for his brother, who was diagnosed by the school psychologist as mentally retarded. There he encountered some problems. His accent and unfamiliarity with the language made it difficult for him to be understood. While the worker talked about residential placement and education, Julio resisted, indicating the family just needed help with the local school, where he felt his brother had been misunderstood because of a language problem. The local school had suggested Julio and his mother go to the case management unit because school officials

---

*Adapted with permission of McGraw-Hill Companies.

did not believe they could provide adequate services. Julio felt their referral indicated insensitivity and an unwillingness to be concerned with keeping the family together and helping his brother function better in English. The vast array of services being offered at the case management unit was bewildering to Julio. He was inclined to simply withdraw from the situation and tell his mother to keep his brother at home.

To the worker at the case management unit, Julio seemed strange. She did not exactly use that word, but she wondered with some exasperation why he did not want to take advantage of the many services available to his brother. Why did he seem so reluctant to keep appointments both with the worker and with the school? Why had he withdrawn his brother from school when this was clearly against the law for a child so young? To the case manager, Julio's behavior was inexplicable.

It is always the majority group that defines who is a stranger and who is not. The people who seem strange to us are not strangers to those with whom they hold common cultural traditions. If the situation were reversed and we found ourselves in a place where our cultural ways and values were different from the majority, then we would be the strangers.

Gudykunst and Kim (1997) use the concept of the stranger to define those whom we encounter who seem strange to us, whose ways of thinking and acting are unfamiliar, and who are not members of our in-groups. In other words, they are people who are close enough that we cannot ignore their presence, but they are unfamiliar to us and therefore seem like strangers. (Throughout this chapter, we will use the term *stranger* as it is defined here.) If people come from another culture, possibly from another country, it is entirely possible that they do not know enough about your culture to be able to relate easily to you. In addition, it is quite likely you do not have enough information about their culture to be able to make the new situation smoother for them.

As the human service professional, you are the one who can take the initiative in making the adjustment smoother for those who come to us as strangers. When people we might consider strangers have developed a good degree of competence in the majority culture and can communicate well, they will be healthier. Studies, however, indicate that it takes a long time for immigrants to adjust to the new culture and that if this maladjustment is severe or long-term, it can cause serious mental health problems as a consequence.

As the world community becomes more global, we can expect to encounter people from many different cultures who will seem like strangers, people who are different. Gudykunst and Kim (1997) make the point that we have internalized our own culture to such a degree that we believe it is innate in some way. They write, "anyone whose behavior is not predictable or is peculiar in any way is strange, improper, irresponsible, or inferior" (p. 357).

When people deviate from the familiar, we are likely to notice it instantly. We may feel anxious or surprised and uncertain. We may be forced to look more closely at our own cultural assumptions. Perhaps we are forced to conclude that aspects of our culture that we have taken for granted are not particularly useful. Our cultural identity may be challenged. Obviously it would be easier to avoid all this and stay away from strangers. Many people do just that, preferring not to experience these unsettling emotions. In human services, however, our work is all about encounters with people, and our purpose is to be helpful. Avoiding strangers cannot be accomplished unless we are irresponsible. For that reason, we need to know what to do when we encounter strangers.

ANXIETY AND UNCERTAINTY

It is common for most people to feel uncertain or anxious when they are attempting to interact with people from other cultures. If we are consumed with our uncomfortable feelings, this will impede our communication with strangers. For this reason, we need to be able to manage our feelings during these encounters to provide for a constructive exchange.

Many times we attempt in some way to reduce anxiety or stress in these encounters. We project our notions about what the person means, giving us a certainty that might not be justified. We might try to develop theories about the other person that feature similarities. The more we believe a person is like us, the less likely we are to feel anxious. Thus, we might look for similar psychological reactions, similar group affiliations, and similar cultural aspects.

When Misaki came from Japan to study at a local college, she was the only person from Japan on campus. Other girls in her dorm invited her to join them for meals and to walk to classes with them, and Misaki did so. When the other girls laughed and talked about trivial matters, Misaki was silent. She rarely made small talk with the girls. They began to interpret her behavior as "too serious" and worked even harder to draw her into their discussions. Misaki was always pleasant but contributed little to these exchanges.

To the girls on her dorm, she seemed too serious; but to the resident assistant, Misaki appeared depressed. Ann, the resident assistant, based her opinion on the fact that Misaki never looked up when she spoke and never looked directly at Ann. Ann asked Misaki if she would like help with her sadness or depression. Misaki said "yes," so Ann made a referral to the campus clinic. There an intake worker decided that Misaki was indeed depressed and concluded it must be about leaving her homeland. To each question the intake worker asked, Misaki answered "yes." Yes, it was hard being here in America; yes, she missed her parents; yes, she had trouble understanding everything the professors said in class.

A counselor at the clinic recommended to Misaki that she join in more with the girls on her dorm and learn to "loosen up and have fun." In Japan, however, people who talk a lot are not viewed as particularly trustworthy. Those who use silence more frequently are considered discreet and trustworthy.

The girls in the dorm, the resident assistant, the intake worker, and the counselor in the clinic did not understand Japanese culture well enough to refrain from judging her by their own culture. Furthermore, in Japan people often assess what it is the speaker wishes to hear and answer "yes" as a means of keeping the social harmony. They might not mean yes in precisely the way an American might interpret it. In addition, Japanese people often do not look directly into the eyes of someone with whom they are conversing. To look directly at another is a sign of defiance or aggression. Looking away is a sign of respect. The Americans took Misaki's behavior as indications of her shyness or depression because that is what the behavior would most likely mean in American culture. By fitting Misaki's behavior into American cultural meanings, the Americans did not have to feel anxious about how to interpret the behavior of a stranger.

In America, girls talk to each other frequently and often about trivial matters as a way of cementing their ties to one another. To the girls who befriended Misaki, this seemed normal. Her silence did not. In order not to feel anxious about her, they projected their own theory about her behavior onto her, that she was sad about leaving her home in Japan. This was a normal reaction, understandable from their standpoint. The theory made Misaki seem more like them, thus their theorizing reduced their anxiety.

We will unwittingly go to great lengths to resolve our anxious or uncertain feelings. Often what we do is inaccurate and not useful in promoting clearer communication.

THOUGHTLESS VERSUS THOUGHTFUL COMMUNICATION

First, we need to find a way to control our anxious feelings to allow us to really be able to listen and communicate. If we are likely to feel anxiety when we talk to strangers and if this anxiety is going to interfere with a realistic understanding of these strangers, we are going to hear and communicate in a skewed or inaccurate manner. What is worse is that we may be only vaguely aware of this problem. The following sections discuss areas you can evaluate to make your communication more thoughtful and accurate.

Recognizing Our Tendency to Categorize

Think about what we do when we are communicating without thought. We categorize people; we assume there is only one correct or normal way to view things; and we are closed to information that does not fit with our cultural perspective. When we encounter someone who is different, we dump that person into one of our large categories or stereotypes. When someone's behavior does not fit or his or her thinking is strange, we are thrown off balance. To prevent that, we have categories all ready into which we can place this person. Because this is probably a habit, we are not completely aware of our categorizing. Actually, you may need to add many new categories that are much more specific and definitive, categories that will be better predictors of behavior.

Looking for Exceptions

To become more thoughtful as you communicate, start to look carefully for the exceptions to your categories. If you think all Hispanics are loud, look for times when you have encountered Hispanic people who were not loud. If you believe all Muslims are militant, look at times when individual Muslims have expressed cooperation. If you believe all Jews horde money, look at times Jewish people have been generous. In other words, recognize that the categories you have probably been using are likely to be much too broad to account for all the specific differences you may encounter in the people you serve.

Another way to look for exceptions to your categories is to seek differences in each specific individual. If you dismiss a stranger as coming from a group that is generally believed to be resistive to your efforts to help, you will have little success compared to recognizing the resistance and then looking for times the stranger was not resistive. If you have categorized someone as too talkative, look for times when that person was listening instead. If you are sure the people in a particular group are stupid, look for times individuals in that group made wise decisions or choices. Seeing others as individuals will make these exceptions important to you. As a competent worker, you will diligently seek these exceptions to gain a more accurate understanding of the person you are serving.

Checking Our Attributions

Most research shows that when we see a stranger's behavior as negative, we are inclined to blame that behavior on the stranger's character or disposition. When we see the stranger's behavior as positive, we are more likely to think this person is an exception and attribute the exceptional behavior to the environment or the circumstances. In other words, it appears that for many of us, giving up our stereotypes is very hard. We would rather see the exceptions to our stereotypes as something external to the stranger, and we are likely to blame behavior that seems to fit the stereotype on the personality of the stranger.

The opposite is often true as we go about attributing causes to our own behavior and the behavior of those we consider to be members of our in-group. If we see negative behavior in these people, we are likely to blame the environment or circumstances, while we often consider positive behavior a reflection of the person's character. The following summarizes how we often see things:

Our positive behavior	Attributable to our good character
Their positive behavior	Attributable to the environment or the circumstances
Our negative behavior	Attributable to the environment or the circumstances
Their negative behavior	Attributable to their poor character

When we do this systematically, it reveals prejudice on our part. In addition, these systematic errors in attribution cannot have come about thoughtfully. They are thoughtless, automatic ways of looking at other people. As you become more thoughtful in your communication, become aware of how you explain the behavior of others.

Evaluating Scripts

We have all learned certain scripts for the activities we engage in frequently. For instance, if you meet someone you see often, but do not really know very well, you might say "Hi" as you pass that person. She might say, "Hi. How are you?" You would probably say something like, "Fine, and you?" She might then respond with "Just fine, thanks." By the time this exchange is completed, you may be several yards apart and walking in opposite directions. This constitutes a script for passing someone you see every day but do not know very well.

There are scripts for a variety of everyday activities. We carry them in our heads to be used when the appropriate situation presents itself. We have learned them first by observation and then from our own participation in these activities. The exchange demonstrated in the previous paragraph did not take much thought. Two people may pass each other every day and go through much the same exchange. They do not stop and consider what to do as each encounter presents itself.

We expect that everyone from our culture will respond to our "hello" with a "hello" of his or her own. If one individual does something different, we are thrown off balance. People from other cultures, however, have learned different scripts. For instance, a common area of misunderstanding relates to the fact that different cultures have different nonverbal ways of indicating that they do not want to be approached. Suppose you are indicating through your body language to someone from another culture that you do not want to be approached, and this person approaches you anyway. You may feel pushed and invaded when, in reality, the stranger could not recognize the signals. You could make a similar mistake. You might see a person you want to join you. You might wave to them and point to the group to indicate they should come over and join you. To a person from an Asian country, this would be insulting. Waving people over, and especially using one finger to do so, is considered rude. We are looking at two different scripts.

Behavior or communication that seems strange to you may simply be a different script presenting itself. Stop and think about what the unexpected behavior means to the stranger. Is this offensive behavior, or does the stranger mean something quite different? Is there a possibility that you are misreading the signals or cannot recognize the signals from this stranger? Is it likely that the signals you are sending are not familiar to the stranger?

Checking Perceptions

Gudykunst and Kim (1997), who have written extensively on this subject, recommend that we simply check our perceptions with strangers to see if these perceptions are accurate. Instead of assuming we know what a stranger means, we need to check.

These authors recommend a three-step process:

1. Describe the other person's behavior, being careful to simply describe what was observed without evaluating or labeling the behavior.
2. Tell the stranger how you interpreted his or her behavior. In doing so, be matter-of-fact. Refrain from any hint of a negative evaluation of the behavior.
3. Ask the stranger if your perceptions are accurate.

Checking your perceptions is a good way to keep the communication between you and the stranger accurate and meaningful. It is important not to assume you know what the stranger means or what the stranger feels. Check with that person to see if what you perceive is correct.

Allowing Differences

It cannot be stressed enough that thoughtful communication is extremely important in reaching real understanding with strangers. Obviously, the better the understanding between you and a stranger, the more likely it is that you will be effective and competent in your assistance to that person.

Not all strangers will respond the same way to their new environment. Differences between the culture of the stranger and the culture of the host society will account for how a stranger responds. Large differences in verbal and nonverbal behavior, in norms or language, or in political and religious orientation can make adapting to the new surroundings more difficult. Where the differences are small, things may be easier for the newcomer. For example, someone from Canada would have less trouble adjusting to the United States than someone from Botswana. When you take this into account, you are able to look more thoughtfully at the stranger's attempts to adapt and be more helpful in that process.

In addition, recognize that there is a lot you do not know and be open to finding out more. When you are communicating with someone who is a stranger to you, put aside your goal for that conversation and begin to listen carefully for new information the person might be providing to you.

Finally, accept that there is more than one way to view something or to understand something. People have different perspectives; but that does not mean that some are superior to others or more correct than others. We may have been taught that this is so, but look at other ways to explain behavior besides your own perspective. Try to understand what perspective the stranger may have. This can be done only if you communicate thoughtfully.

DIMENSIONS OF CULTURE

Researchers in the field of communication have looked for ways to help us understand cultural differences even when we do not know the details of every culture. They have proposed that cultures have an underlying foundation of individualism or a foundation of collectivism. Cultures fall along a continuum, with no single culture being all one or the other; but many researchers believe that communication can be facilitated between people of different cultures if we know whether the stranger with whom we are communicating is from a culture that is primarily an individualistic culture or a collectivistic one. This tool is particularly helpful when we do not know all the particulars of a specific culture.

Individualistic and Collectivistic Cultures

Using information from Gudykunst and Kim (1997), we will look at some of the characteristics of cultures that are predominantly individualistic or collectivistic. Figure 4.1 lists some countries and shows how they fit into this classification scheme.

How Individualistic and Collectivistic Cultures Differ

First, individualistic cultures tend to place a higher value on the individual than on the group. Collectivistic cultures, on the other hand, tend to place more value on the group.

Tend to Be Individualistic Cultures *(based on predominant tendencies in the culture)*	Tend to Be Collectivistic Cultures *(based on predominant tendencies in the culture)*
Australia	Brazil
Belgium	China
Canada	Columbia
Denmark	Egypt
Finland	Greece
France	India
Germany	Japan
Great Britain	Kenya
Ireland	Korea
Italy	Mexico
Netherlands	Nigeria
New Zealand	Panama
Norway	Pakistan
South Africa	Peru
Sweden	Saudi Arabia
Switzerland	Thailand
United States	Venezuela
	Vietnam

Figure 4.1 Individualistic and collectivistic cultures

Individualistic Cultures

Individual More Important

- Individuals should look out for themselves and their families
- Promote self-fulfillment
- Emphasize individual initiative and achievement
- The in-group influence is very specific to times and place
- Individual goals are emphasized
- Tend to apply their value standards to everybody (universalistic)
- Emphasize needs and goals of the individual over the group
- Support unique individual beliefs

Collectivistic Cultures

Group More Important

- Members of in-groups look out for each other in exchange for loyalty
- Require that people fit into the group
- Emphasize belonging to groups
- The in-group influence is very general over all situations
- Group goals are emphasized
- Tend to apply different value standards to members of their in-groups and to members of out-groups
- Emphasize the needs and goals of the group over the individual
- Shared in-group beliefs

Another difference lies in the way in which society is viewed. In individualistic cultures, there is ranking and hierarchy, while collectivistic societies tend to be more equalitarian.

Individualisitc Cultures

Have a More Vertical Culture

- People are expected to stand out from others
- Value is placed on freedom
- Maximizing of individual outcomes

Collectivistic Cultures

Have a More Horizontal Culture

- People are not expected to stand out from others
- Value is placed on equality
- Cooperation with in-group members

There is also a difference in the way the two cultural types use the surrounding context in communicating. In individualistic societies, the communication tends to be so direct that a person rarely needs to check the context to fully understand the meaning. In more collectivistic societies, context is extremely important.

Low-Context Communication

Context Is Only Minimally Important

- Tend to use very direct communication
- Listener does not have to use context to obtain meaning
- Communication is less ambiguous
- More concerned with clarity as necessary for effective communication
- Communication is about the same for in-groups and out-groups
- Value saying what you think
- Value truthfulness
- Verbally direct, precise, and absolute
- "Yes" means agreement

High-Context Communication

Context Is Important in Determining Meaning

- Tend to use more indirect communication
- Listener must use context to obtain meaning
- Communication is more ambiguous
- More concerned with avoiding hurting others or imposing on others
- Communication is very different for in-groups and out-groups
- Value avoiding confrontations
- Value courtesy
- Verbally indirect, imprecise, and probabilistic
- "Yes" does not necessarily mean agreement

Looking, therefore, at major differences in communication between the two cultural types, we find the following:

Individualistic Cultures

■ Low-context communication is more precise
■ Direct and explicit

Collectivistic Cultures

■ High-context communication uses understatements, pauses, silences, or a shortage of information
■ Indirect and implicit

As these comparisons indicate, there is plenty of room for misunderstanding. A person from a culture that values clear, explicit information might suspect someone from a high-context culture of being manipulative or confused. Someone from a horizontal culture might find someone from a vertical culture rude and boorish or incredibly selfish. If a client from a collectivistic culture waited to engage in services until he had group consensus, the worker from an individualistic culture might mistakenly think the client was resisting treatment or uninterested in help. If a worker from an individualistic culture encouraged a woman from a more collectivistic culture to look out for herself and leave an abusive marriage, the client might feel helpless and unsupported. Leaving the group might not be an option for her.

One common error is for people from individualistic cultures to assume the person with whom they are speaking from a collectivistic culture is speaking as directly and explicitly as they are. Individuals from collectivistic cultures can make the reverse mistake, assuming the person from an individualistic culture is only implying or speaking indirectly.

Privacy and Self-Disclosure

At different times, we feel open to interaction with other people or we feel closed and seek privacy. Different cultures regulate privacy needs in different ways. While individualistic cultures do so with physical boundaries, collectivistic cultures do so by psychological means. For instance, in collectivistic societies, people who might be encountered in general public situations are often seen and treated as nonpersons and simply ignored. In this way, the individual is protected from unwanted involvement. In individualistic societies, this would be seen as rude.

Individualistic Cultures

Privacy Regulation

■ Use of physical barriers such as doors, walls, private rooms and offices, fences, hedges

Collectivistic Cultures

Privacy Regulation

■ Use of psychological barriers such as speaking softly, treating one another with decorum, treating people in public as nonpersons, sending nonverbal cues that approach is not desired

Self-Disclosure

■ More likely to self-disclose because privacy is protected through physical barriers

Self-Disclosure

■ Less likely to self-disclose to protect accessibility of others

Time

Time is conceptualized differently in different cultures. How people conceive of time will determine how they are likely to use it as well.

Individualistic Cultures

Monochronic Time

- Time is seen in discrete compartments
- Compartments are used to schedule events one after the other
- Actual time on the clock is more important
- Emphasize adherence to schedule
- Punctuality important

Collectivistic Cultures

Polychronic Time

- There are no compartments
- Can do more than one activity at a time
- Activities are more important than the time
- Emphasize completion of tasks
- Punctuality not so important

Face

It is common for people in individualistic cultures to talk about saving face. In sociological terms, *face* refers to a public self-image. In collectivistic societies, there is an emphasis on protecting the face of others, a concept that is less emphasized in individualistic cultures.

Individualistic Cultures

- Less emphasis on respect for elders and superiors
- Concern with saving one's own face

Collectivistic Cultures

- Emphasis on respecting or giving face to one's elders or superiors
- Concern with saving the other's face

Persuasion

Different cultures use different methods for persuading people to undertake certain activities or to comply with specific requests.

Individualistic Cultures

- Focus on the person they are trying to persuade
- Direct requests for what is desired
- Likely to threaten the person's security
- Likely to state negative consequences to the person if he or she does not . . .
- May ingratiate themselves to the other (I really value you, therefore . . .)

Collectivistic Cultures

- Focus on the context in which the persuasion is taking place
- May use altruistic strategies (for the sake of the group, company, and so on)
- Likely to appeal to duty, concern for the whole group
- More likely to promise positive consequences if he or she does . .
- May imply a "good" person would do this

Expression of Emotion

Different cultures express emotions differently and use the display of emotions to further the cultural values.

Individualistic Cultures

- Less concerned with using emotions to further group cohesion or harmony
- Display more of a variety of emotions
- More likely to express positive emotions with members of out-groups

Collectivistic Cultures

- Display those emotions more likely to support group cooperation
- Not as tolerant of free expression of a variety of emotions
- Negative emotions toward others are expressed privately so as not to reflect badly on the in-group
- Negative reactions to members of the in-group are withheld so as not to disturb the harmony of the group
- Negative reactions to members of the out-group are more often expressed to increase the cohesion of the in-group

Information Seeking

In all cultures, people attempt to gather information that will clarify situations and reduce anxiety. Members of individualistic and collectivistic cultures go about this task differently.

Individualistic Cultures

- Try to get to know the person, such as characteristics, beliefs, past experiences, attitudes
- Looking for personal similarities with the member of an out-group
- Tend to self-disclose to strangers

Collectivistic Cultures

- Try to get to know the person's group affiliations, age, and status groups
- Looking for group similarities with the out-group member
- Tend not to self-disclose to strangers

Interestingly, research indicates that in America, European Americans tend to self-disclose more than African Americans do. When close friendships are formed, the opposite is true. In close relationships, African Americans will self-disclose more than will European Americans.

Conflict

Look at the differences among cultures in dealing with conflict.

Individualistic Cultures

- Prefer to deal with conflict directly
- Not too concerned that all parties save face
- May look for ways to integrate conflicting views or compromise

Collectivistic Cultures

- Prefer to deal with conflict indirectly
- Concerned that all parties save face
- May try to avoid the conflict or give in to the other

Research suggests that "Chinese prefer bargaining and mediation more than North Americans, . . . Mexicans tend to avoid or deny that conflict exists, . . . Canadians prefer negotiation, . . . Nigerians prefer threats more than do Canadians" (Gudykunst & Kim, 1997, p. 282). In the United States, people are more likely to deal directly with conflict, looking openly for ways to resolve it. With all of these cultural differences in the preferences for handling conflict, it is important to approach conflict thoughtfully.

OBSTACLES TO UNDERSTANDING

There are any number of mental mechanisms we use that can block clear communication. These mental mechanisms are used primarily to reduce anxiety when we encounter strangers. Often we are not aware of the extent to which we resort to these mechanisms. They are particularly present when we are engaging in thoughtless communication. The following sections discuss some of the obstacles that prevent real understanding.

Stereotypes

Some stereotypes are positive, and some stereotypes are held only loosely. Communication is most likely to be obstructed by rigidly held, negative stereotypes that can lead a person to make inaccurate assumptions and predictions about another person. Sometimes a person fits our stereotype of a particular group, but many other persons in the same group may not fit the stereotype at all. Becoming aware of our assumptions and questioning them is important.

Ethnocentrism

Ethnocentrism means that we use the standards common in our own culture to judge the behavior and culture of other people. It is important to understand that we are all ethnocentric to some extent. It is common to look at others through the lens of our own culture. The way our culture is arranged seems "normal" or "correct." Deviations from our culture, therefore, seem abnormal and incorrect. It is not that we consciously decide to employ ethnocentric tactics, but rather we are socialized into viewing the world in a particular way.

The antidote to ethnocentrism is cultural relativism. Using cultural relativism, we try to understand the meaning of others' behavior and communication within the context of their culture, not our own. When we use ethnocentrism to judge people from other cultures, we create barriers and distance. When we use cultural relativism to understand others, we diminish barriers and distance.

Prejudice

Gudykunst and Kim (1997) define prejudice as "judgment based on previous decisions and experiences" (p. 124). People hold prejudices against whole groups of people and against individual members of those groups when they are encountered.

Prejudice involves an attitude that generally stems from a negative stereotype. If you think all the people in a particular group are pushy and devious, you will probably decide that you do not like those people. Not liking those people is a prejudiced attitude that is based on the stereotype you hold in your head about members of that group. This easily leads to discrimination, in which you take pains to avoid being around these people whom you do not like. You might deny a member of this group a job for which he is qualified or seek to deny a family from this group housing in your neighborhood.

Conflict

When we are already suspicious or uncomfortable with a group of people, misunderstandings can turn into hostility and conflict very easily, particularly since we are likely to attribute the negative behavior of strangers to their personal characteristics, while attributing the negative behavior of in-group members to the situation. All of the mental mechanisms described in the previous sections serve to make other groups seem less worthy of being understood and enhance the possibility that conflict will occur with individual members of a particular out-group.

Be aware of two points. First, misunderstandings may stem from mental mechanisms that are inaccurate or have obstructed real understanding on everyone's part. Second, once a conflict has occurred, the approach to resolution may be quite different from one culture to another.

Changing Attitudes

It appears from recent research that we can change our attitudes toward strangers or members of out-groups through a number of opportunities to interact with them positively. Stephan (1985) talks about ways to increase understanding and to create more favorable relationships among groups. For instance, an emphasis on cooperation, rather than on competition, is helpful. It is useful if those coming together have about the same status within their own groups and have some similarities in common with each other. Supporting the individuality of each member helps smooth things out. Voluntary contact and contact that is focused on substantive issues as opposed to superficial issues are more useful. Everyone involved should work toward a positive outcome.

The goal of intergroup cooperation and contact is to learn to see members of other groups as individuals rather than as representatives of our own biases and stereotypes. Gudykunst and Kim (1997) address this in their dichotomy between uncertainty oriented people and certainty oriented individuals. They write about an uncertainty orientation:

> Uncertainty oriented people integrate new and old ideas and change their belief system accordingly. They evaluate new ideas and thought on their own merit and do not necessarily compare them with others. Uncertainty oriented people want to understand themselves and their environment. (p. 185)

A certainty orientation is quite the opposite. The authors write:

> Certainty oriented people, in contrast, like to hold on to traditional beliefs and have a tendency to reject ideas that are different. Certainty oriented people maintain a sense of self by not examining themselves or their behavior. (p. 185)

When we are communicating thoughtfully, we can make a conscious choice to acquire more of an uncertainty orientation toward new situations and strangers. By remaining open, we can learn more.

We need to go one step further, however, by offering confirmation to others with whom we communicate. When you are working with people who are strangers to you, confirm for those people that they are valuable to you as individuals, that their experiences and concerns are important, and that you are willing to become involved in helping them to resolve their problems. When we deny that another person's concerns and experiences are valid and therefore imply that they are insignificant, we demean that person as an individual, and the opportunity for meaningful resolution and rapport is lost.

COMPETENCE

It is clear that those workers who are adaptable to the situation and flexible in choosing how to respond to a situation do better in cross-cultural communication. This means that the workers are intuitive and sensitive to what others might mean or need and open to considering the interaction from a number of different points of view. Figure 4.2 highlights some of the points that might make this process easier.

Competence in cross-cultural interactions depends very much on the individual worker's commitment to give high-quality service to every person who comes for assistance. As you begin to practice, you will encounter more people from specific minority groups or cultural groups with which you are unfamiliar. Ethically you are responsible for developing an understanding of their culture or subculture, at least along the various dimensions we have outlined here.

Until that study has been completed, it is important to consciously and thoughtfully monitor your interaction with strangers. Make certain that you hear the significance of their concerns and experiences, that you respond in a way that lets them know they have been heard, and that you provide a respectful environment where problems can be resolved.

TESTING YOUR CULTURAL COMPETENCE

Instructions: Look at the culture of each client described in the following scenarios, and decide what might be the underlying issue. What are you thinking about the client as you read each description? What are your first theories about what constitutes the client's problem? Do you have a problem personally with the behavior of the client, and if so, in what way?

These brief explanations may help you answer the questions when you read the scenarios that follow:

- In many Asian cultures, members do not talk about family problems, feeling that these are private. They may also pretend that no problems exist.
- In most Asian cultures, crossing the legs and pointing the toe at another person is considered extremely rude. Because members of Asian cultures like to maintain harmony, they would not be likely to tell you directly that they were offended. In addition, in most Asian cultures, waving at another person or indicating that a person should join you by calling them over with your hand or a finger is also considered extremely rude.
- Asians are not likely to make changes in the family or to engage in discussions about the family unless the male head of the household is present or is consulted. Furthermore, they are likely to tell you things are all right in order to maintain harmony. Things may not be all right. They may also tell you that you were helpful to them because they assume that is what you want to hear, not because it is true. Telling you what they believe you want to hear will maintain harmony.
- In most Asian cultures, group needs and considerations are more important than individual needs and considerations.
- In most Hispanic families, the man makes the major decisions and expects to be consulted about anything affecting the family. He would not be likely to

For Individualists to Interact Effectively with Collectivists, Recognize That for Collectivists

- Emphasis is on group membership.

- Group membership can be used to predict the collectivists' behavior.

- When group membership changes, the collectivist behavior changes.

- Unique relationships, where one has more status than another, are comfortable.

- Competition is seen as threatening.

- Harmony and cooperation are emphasized.

- Keeping a positive public self-image (face) is a concern, and individualists can help them to maintain that.

- Criticism and the person being criticized are not seen as separate, thus avoiding open confrontation is best.

- Deliberately cultivated long-term relationships work better.

- More formal initial interactions are preferred.

- Forced self-disclosure will cause a negative response.

- Respect for age and position is important.

For Collectivists to Interact Effectively with Individualists, Recognize That for Individualists

- There is a degree of emotional attachment from their in-groups.

- Behavior cannot be accurately predicted based on group membership.

- Out-groups are not seen as extremely different from in-groups.

- Equal relationships where status is eqaul are preferred.

- There will be pride in their own accomplishments.

- Arguments that emphasize harmony and cooperation are not very persuasive.

- Saying negative things about others is more likely.

- It is easier to separate criticism from the person being criticized.

- Long-term relationships happen less often.

- Relationships that contain more rewards than costs are more likely to be maintained.

- Initial friendliness is not considered a sign of an intimate relationship.

- Respect based on age, position, or sex is generally not as likely.

Figure 4.2 Points to remember in cross-cultural interactions

take his wife's ideas or concerns into consideration. She would be expected to defer to the husband.
- Many Hispanic families allow mental health problems to exist for a very long time, rather than admit that there is a psychiatric problem.
- In Hispanic culture, it is often believed that depression is due to a lack of religious faith.
- In Hispanic culture, mental health problems are often attributed to sin.

1. A man from Vietnam is in your office because his 11-year-old daughter has been having trouble in school. The school suggested the daughter be tested by your agency. You are doing the intake, but only the father has come into the office. He is

very reluctant to tell you any specifics, but talks instead in extreme generalities. What might be the source of his reluctance to talk to you in detail about his daughter's problems?

2. A Japanese family in the emergency room is seeing you because of a serious accident in which their teenage son was severely injured. During the course of the conversation, you cross your legs so as to be more comfortable. The family continues to talk to you in a polite but superficial manner, and gradually each member drifts away—to get a soda, to use the restroom, and so on. They are obviously resistant to sitting down with you again.

3. A Chinese woman is hospitalized for a serious infection, and her doctors think she seems depressed over possible home problems. You talk to her, and she appears to reassure you that everything at home is fine. When you come in the next time, she wants you to talk to her husband instead. He, too, is reassuring and pleasant. Later you ask the woman if your talking to her before was helpful; and she smiles and tells you it was. You are not sure.

4. You work in the school counseling office and you have been asked to help a gifted young Vietnamese student fill out applications to several prestigious colleges. She is in line for a number of scholarships. She works on the applications, but with obvious reluctance, and indicates she cannot consider going away to school until the family has decided what she will do.

5. A Puerto Rican woman is referred for depression by her family doctor after the birth of her fourth child. At the intake interview, her husband comes and answers all questions. He indicates that he will decide what she needs and what is to be done. The wife says very little and speaks only when she appears to feel her husband wants her to do so.

6. A Mexican-American family brings an elderly aunt to the emergency room. The older woman is severely depressed and emaciated. She also appears to be nearly catatonic. She is admitted immediately to the hospital on an emergency commitment. You learn that this woman has been in severe depression for years and

wonder why the family waited until things became so serious. Later when you talk to the aunt, she is somewhat improved. She indicates that her pastor visited her and told her that her depression is due to her lack of religious faith. She tells you she agrees with this assessment. She tells you she believes that if her faith were stronger and she was less sinful, she would not feel like this.

7. A young Hispanic woman is admitted to the hospital, and the doctor believes she is showing obvious signs of schizophrenia. She is hallucinating and has not been eating. Her family tells you this problem is the direct result of her sinful behavior. According to them, she is too friendly with the boys in her class. The students often call each other to compare homework, and sometimes a group of boys and girls will go to a party or several boys and girls will walk home from school together. The family has tried to get her to cut off these friendships and to stay home and help her mother more. Because she did not listen, she is being punished.

Chapter 5

Examining Attitudes and Perceptions

INTRODUCTION

Attitudes are extremely important. The feelings you have about other people are bound to be communicated to those people one way or another. If your attitudes are positive and supportive, you will establish rapport. If you feel superior or disdainful, no matter how well you try to hide those feelings, they will eventually be communicated to your client and you will lose a working relationship.

Good human service workers have learned about themselves—their fears, sensitivities, and errors in judgment. In facing those things about themselves, these workers have come to understand themselves in a way that allows them to feel understanding and warmth toward themselves and others. If you are able to forgive yourself for the mistakes you make and the struggles you have had and see them as an important part of growing, you will recognize that you are basically all right. It is then much easier for you to understand others who are making mistakes and who are struggling with issues and problems. You know that the problems you have faced have provided valuable lessons. These personal struggles have helped you to grow into a more sensitive and insightful person. Problems and unfortunate decisions happen to everyone. You will, therefore, be more inclined to see personal struggles as productive of growth, and not necessarily a reflection of inadequacy.

Begin, therefore, with yourself. Be tolerant of your mistakes. Look at yourself objectively. Forgive yourself for errors in judgment, particularly errors that have taught you important lessons or helped you to grow. Recognize that part of being human is to struggle with issues and transitional problems and that through this process, we are often strengthened and given new insight.

BASIC HELPING ATTITUDES

There are three basic helping attitudes you will need to bring to your work. Studies have shown that even in the absence of much formal training, workers who genuinely care for their clients and are committed to them will be able to help their clients make important changes and move toward better circumstances and increased emotional health.

Warmth

A worker needs to be friendly, nonjudgmental, and receptive. These three attitudes create a warm atmosphere, one that serves to put the client at ease.

In your presence, clients feel valued by you as a person. You communicate a belief to them that they are worthy of being understood. You do this through your actions, body language, and the way you listen to each person. In addition, you refrain from evaluating what clients say and the actions that they take. For instance, a warm person would say something like, "Tell me more about that." A judgmental person might be more likely to say, "Oh well, you should never have done something so careless."

In addition, you are receptive to what people have to say. You listen to what they tell you they have done and felt, what brought them to seek your assistance in the first place. When you say, "Tell me more about that," you are inviting the person to open up and talk safely with you. When you say, "Oh well, you should never have done that," you have passed judgment and cut off further discussion. The atmosphere is no longer warm.

Part of being warm is not dominating clients. It means you respect your clients' right to make their own decisions. You may facilitate better decisions than the ones they might have made alone, but ultimately the decisions about their life are their own, and you will respect that. Sometimes we are painfully aware that a decision is not the right one for the person to pursue. Later in this book we will talk about how you can give the client some of your thoughts in ways that the person can hear and use your ideas more easily. Nevertheless, a person may choose to ignore you and make her or his own decisions, and you will respect that. We have all made unfortunate decisions and concluded, as part of our growing process, that they were not very useful.

Some people think that in order to be a warm person, they must stand by passively, never confronting or giving the client another way of looking at things. A warm person is still able to facilitate change, and change is often painful. As you learn the skills in this course, you will pick up methods to use that help clients see things from more than one perspective and possibly see characteristics about themselves that are difficult to face.

Genuineness

You have heard the expression "be yourself" many times. This is a must for those who help others. Nobody relates well to a phony. People sense when you are not being authentic. Perhaps you use slang that sounds forced, phrases you do not usually use. Maybe you are using them now to make the client think you are familiar with his lifestyle or culture.

Maybe you put on a phony dialect or use profanity to seem more down to earth. You might pretend to be a physician, wearing a white coat and allowing your clients to call you "doctor." The client calls you "doctor," and you do not correct the impression. You might pretend to have degrees you do not have. You could use big words you know the client does not understand. None of this is authentic. You will be seen as a person who is rather foolish at best, and untrustworthy at worst.

Be open and truthful. Strive to be yourself, to present your authentic self to the other person. If you do not know something, say so. If you lost your client's forms, tell her that and apologize. If your client asks about your credentials, tell him matter-of-factly what they are.

Empathy

Empathy means putting yourself in the other person's shoes. To do this you must be able to comprehend what that person's needs and feelings are. In human services, we often say that we are able to listen with the third ear. We hear more than what the client is telling us. We hear the underlying emotions, desires, and worries. With practice we become proficient at this, and we develop a special sensitivity.

Part of empathy is being able to accurately communicate to the person an understanding of these underlying emotions. If you are able to put into words the feeling the person is experiencing right now, you are practicing empathy. To do it well you will communicate in a way the other person can understand and accept, not in a way that is threatening or judgmental. An empathetic person would say, "You must have felt sick when you saw what the accident did to your car." Someone with little empathy would be more likely to say, "I'll bet you could have shot yourself when you saw what you did to the car. Do you have insurance? Bet you didn't have that either."

Finally, empathy is not sympathy. It does not mean that you are so sad for the person that you take their situation home and fret about it when you are away from work. It does not mean that you feel sorry for the person and communicate the belief that the other person is a poor soul or the situation is without hope. Sympathy is what we often have for our friends and relations. Empathy is assessing where the client is at any given moment and being able to express that and support it. Sympathy is the common feeling we have for others in pain. Empathy is a basic clinical strategy for supporting people through difficult times.

Minor Problems

The following examples demonstrate some of the minor problems you might encounter with your attitudes.

Minor Problem #1   You assess the feelings incorrectly. For instance, you might misinterpret the client's underlying feelings and thoughts. You might say, "You must have felt sick when you saw what the accident did to your car." The client might respond by telling you she really did not feel sick. Instead, she was angry—furious, in fact—at the other driver who stood there and screamed at her. It is always possible that you may be corrected by the client in this way. This is positive. It allows the client to make you aware of what she is really feeling and thinking, bringing you much greater clarity.

Minor Problem #2   Your mind wanders. There may also be times when you are not listening attentively. Something has happened at home; you just finished handling a personal problem; or you hoped to go home before 4:00 and now it is 5:00 and this may take some time. Your mind leaves the immediate situation and wanders to your personal concerns. Of course you do not want this to happen often, but it will happen. If you are practicing good body language, sitting in a way that communicates interest and attention to the client, these momentary shifts in your focus will not be damaging to your relationship.

Major Problems

The following are examples of major problems that can occur that are not at all useful.

Major Problem #1   You cannot wait to pass judgment. Sometimes workers listen to the client, but their minds are full of judgments they want to make about what the client has said. Rather than truly listening, the worker is judging the client in regard to how well the client handled the situation, whether the client was smart or stupid, whether the client was on top of things or lax in taking care of the situation. The worker cannot wait for the client to be quiet so he can give an authoritative judgment. Rather than listening, the worker is preoccupied with what he plans to say in response. This is unprofessional listening. We might do this with our friends and relations, but it has no place in a professional relationship.

Major Problem #2 You ignore the client's feelings. Another way to miss the important issues is to focus entirely on content and never hear the meaning this situation has for the client. Instead of commenting, "How difficult it must have been," the worker goes on about the actual details. "You mean you went to the corner of Fourth and Market, and this man came from—what—the east, I guess, and asked you directions to the capitol? And then you stepped off the curb toward the car?" There is a place for this kind of listening, but when you do this exclusively and never talk about feelings, important opportunities are missed for healing and building rapport.

Major Problem #3 You cannot wait to offer the solution, to give advice. Important opportunities are lost for rapport and understanding if the worker does not acknowledge the feelings a client is expressing. Some workers rush right past the feelings to the solutions. Rather than saying, "You must have found that painful," the worker says something like, "Well, you'll need to see a lawyer about this. There is a good one around the corner that we use a lot, and I can probably get you in to see her in the next day or so." Anyone can tell a client he needs a good lawyer. Only a very good listener is able to respond with empathy to the underlying feelings present in the client's story.

On Being Judgmental

José, returning from a home visit in a neighborhood where many of the homes were run down and many of the residents were poor, was aghast. "I'm not going back *there*. That's a terrible place to live! Those people should think about how they live and what they are doing to their children." He continued in this way for several days. Eventually the supervisor had an opportunity to talk to him about his attitude and how judgmental it seemed. She explained to him that this is where his clients live and that service requires us to respect the home of another person and not judge it as inferior. Assuming the people who live in the neighborhood are responsible for the way the neighborhood looks and for the people who live there may not be altogether true. "You are not better than the clients you serve," she pointed out to him. "You may feel uncomfortable there, but you can't force a client to come to you here in the office just because you don't approve of the neighborhood in which he lives or because you feel uncomfortable there and don't want to go there."

There are human service workers who sit in judgment of clients applying their own standards to people who are sick, who have been through trauma, or who have no ideas about alternative ways to approach their lives. There are people who say they are in human services to be helpful, but who are actually wary and distrustful of their clients. This is an attitude that has no place in the helping relationship.

For example, Rose, a new worker, was part of a planning team for a new shelter that was about to open for children who needed a place to stay before foster care could be found. These were children who were removed from their homes for physical or sexual abuse or extreme neglect.

During the planning, it was decided that the children would earn points for good behavior and lose points when they violated policies of the residence. Children with ten points or more would get special privileges, while those with no points would stay in their rooms temporarily and miss special activities. The director asked the group, "Where shall we start a child when he or she comes in new? Shall we start them with ten points or none?" Rose was convinced that each new child should start with zero points and earn them. "How about starting each child with ten points and let he or she work to keep them?" the director suggested. While this is what the planning group ultimately did, Rose was upset, thinking that the children would "just come in here and take advantage of us." "They need to know we mean business right from the beginning," she asserted. The assumption that the children would automatically misbehave and take advantage of the staff was judgmental; adding that judgment to the problems the children were already experiencing would be cruel. Rose's need to curtail and punish before any misbehavior had taken place would have put her negative attitudes into action in a destructive way.

There are human service workers who say they have gone into this work to help other people. The minute a client behaves in a difficult manner, however, they complain that they should not have to deal with this sort of person. Rita refused to sit down with the mother of a child who was in treatment and work on a solution to the child's behavior problems. Rita complained that the mother was a difficult person with whom to work. "She's always complaining about what we're doing. She thinks she has a better way. I just want to say to her, 'Look, if your way was so hot, you wouldn't have the problems you've got with Benny.' All she ever does is make stupid suggestions. I just wasn't going to get into that with her."

In human services, we are trained to deal with people who seem dissatisfied and upset. We expect to meet people who do not see things our way, who are challenging, or who question our decisions. Many of the people we work for will have an inaccurate perception of reality or difficulty expressing what concerns them, seem unduly sensitive, or be confused about following a plan. If we only want to work with people we like and who agree with us, people who give us no trouble, we will be barely effective and more likely harmful.

DISCOURAGEMENT

Many of the people we see are discouraged as a result of their circumstances or what has happened to them. Here is a list of symptoms commonly seen in discouraged people:

The situation seems to overwhelm them.

They have low regard for their capabilities or put themselves down.

They are unwilling or unable to take responsibility.

They describe their circumstances or other people as totally domineering and overwhelming.

They have no trouble discussing their problems, but are unable to focus on solutions to those problems.

Their goals seem impossible or unrealistic.

They set impossible standards for themselves.

They sprinkle their conversation with negative global statements such as "*Everybody* hates me" or "*Nobody* ever calls here" or "*Everything* is rotten" or "Bad things *always* happen to me."

WAYS TO MOTIVATE AND ENCOURAGE

Good case managers can motivate clients to reach beyond their former situations. They can encourage their clients to look for and try alternative ways of doing things. Much of this is frightening to people. Change and uncertainty are never easy. Clients, like all of us, resist change. The old ways, however painful or inconvenient, seem preferable to the unknown results of change. The following sections discuss techniques case managers use to help clients make these important changes.

Begin Where the Client Is

An encouraging case manager will accept clients exactly where they are. The worker does not denigrate the client for needing help or coerce the client to be better or work harder.

Arnold R. Beisser (1970), writing in *Science and Behavior Books*, talks about Frederick Perls, the father of Gestalt therapy, and what Beisser calls Perls's "paradoxical theory of change." Beisser defines the theory as this: "Change occurs when one becomes what he is, not when he tries to become what he is not." Perls believed that forcing people to be different never accomplished real or lasting change. First people have to come to terms with who and where they are at the moment and find some degree of acceptance in that. If you start immediately telling clients how to do things differently, how their behavior has been the root of the problem, or how erroneous their thinking has been, they will find it important to argue and defend themselves—to explain why they have done those things or thought that way. Valuable time is wasted and rapport lost as the clients move into a defensive position rather than a collaborative one with you.

As a case manager, you may be the first person to really listen to the clients entering the system. Use this opportunity to bring acceptance and understanding to the way things are now. Let the clients be exactly where they are at the moment they first see you, without remonstrances from you. Beisser (1970) writes, "The premise is that one must stand in one place in order to have firm footing to move and that it is difficult or impossible to move without that footing." To really provide useful assistance to clients and encourage them to grow into their true potential, you must start exactly where they are, not where you think they should be.

See the Client's Strengths

Case managers who encourage their clients will see clients as basically capable and wanting to take as much responsibility for their lives as possible. Start to view people in that way. Clients come with varying degrees of independence and abilities. Some can grow into full independence, and others will always need some help. In this wide spectrum of abilities and needs, a good case manager will see the strengths of each individual client and work hard to help the client maximize and take pride in those strengths.

Accurately Assess the Client's Reluctance

Sometimes clients appear overwhelmed and either unable or unwilling to take responsibility. Try to understand accurately how real these obstacles are for the clients. For example, an inability to use public transportation may be a small obstacle that is easily overcome for a depressed college graduate, but a very real obstacle for an individual with moderate developmental disabilities.

A client may not follow through on projects or goals the two of you set together. The reason may have more to do with the projects than with the client. It may be that the two of you set a goal that is too complicated for now. Perhaps the goal looked safe when the client sat with you in the office, but now it looks terrifying. Explore this with your client. Do not assume clients who do not follow through are being obstinate or uncooperative.

Collaborate

Collaboration is an extremely important skill for all case managers. Throughout this course your instructor will be looking for evidence that you are able to collaborate with your client and that you understand what collaboration means. When we encourage people to try something or move away from a difficult or unhealthy situation, we need to work with them in collaboration to find the best way for the client to do this. It becomes a joint project rather than you telling the client what to do. Talk to the client about what "we" can do or how "we" can look at this differently. Ask the client if she would be willing to consider certain approaches or solutions, and ask what solutions she may have thought of. Give your client an opportunity to express her views or opinions and suggest ways the two of you can work together to overcome the problems.

Appreciate Every Effort

Suppose your client resolves to try something new, and the first efforts are not very successful. To encourage this person to try again, focus on the efforts, not on the results. The attempt to grow or to change is more important to you than whether or not it worked. Point out that the attempt gave valuable experience or information and that it creates a basis for improving.

Never Lose Sight of Potential

As the worker, you will want to focus on the clients' potential. You know all about their deficits or weaknesses. So do they, but they have far less certainty about their potential for growth and change. Look at clients' strengths, their past experiences, and their accomplishments, however small. Use these to guide you in planning the small steps they can take toward positive change.

Encouragement Rules

There are some rules you should follow in encouraging other people:

Start right where the client is.
Use acceptance and empathy.
Work at your client's pace.
Appreciate the significance of all the person's efforts and attempts.
Recognize the client's strengths verbally with positive feedback.
Show confidence in the client.
Keep your sense of humor.

There are some discouraging things you should not do:

Never set a goal for the client without the client's participation.

Never try to force the client to do something he or she really is not ready to try.

Never denigrate the client for not moving faster.

Never demean a client who did not choose your suggestions for handling things.

Never try to get action by shaming the client or making comparisons to others.

Figure 5.1 summarizes the differences between a human services professional who provides encouragement and one who provides discouragement.

HOW CLIENTS ARE DISCOURAGED

There are many ways to discourage another person. You could set up a competition comparing the client to others or to yourself. You could push, force, or shame the client into moving toward some goal. You could spend an inordinate amount of time focusing on the client's mistakes or demand that the person do more or try harder. You could insist that clients do things your way, dominate clients by taking over or demanding perfection or unrealistic outcomes.

You could treat clients like poor souls—as incompetent, bumbling people who need you to do everything for them. You could be insensitive by failing to notice positive changes or ignoring the good things clients accomplish. You could refrain from ever giving feedback except the most negative type. All of these responses discourage others from trying to do more.

Today, many individuals, some of whom hold degrees in fields unrelated to human services, are being asked to give direct care to clients who seem odd and unusual to them. For this reason, the clients may be frightening to these workers. In order to counteract feelings of uncertainty, such a worker may become extremely dominating and coercive. This gives a false sense of control. The domination and need to order clients'

The Encourager:	The Discourager:
1. Says you can	1. Says you can't
2. Helps clients to do their best	2. Wants clients to compete
3. Helps clients look for the best in life	3. Tells clients life is unfair
4. Believes others are capable	4. Believes he or she is more capable than the person he or she is helping
5. Talks directly to people	5. Talks down to people
6. Is supportive of others	6. Is critical of others
7. Is enthusiastic about others	7. Is reluctant and wary of others
8. Is interested in the smallest accomplishments	8. Is interested in the smallest mistakes
9. Helps clients set realistic expectations	9. Has unrealistic expectations for clients and gives up
10. Sees resources and strengths in clients	10. Sees mainly weaknesses and incompetence in clients
11. Lets clients develop their own personal standards	11. Sets personal standards for clients
12. Believes people can improve	12. Believes people never change

Source: Based on *The Encouragement Book* (Dinkmoyer & Losoncy).

Figure 5.1 Encouragement versus discouragement

lives in inappropriate ways are discouraging to clients who are learning to take charge of their lives and make decisions.

An example of this occurred in a program for individuals recently released from an institution for those with mental retardation. A case manager working with the clients visited them every other day in their apartment, where she offered support to help them remain in the community. She enjoyed her work with the clients, and she had received good human service training. When the clients began to make jokes about the pounds they had gained over the years in the institution, she suggested they might want to exercise. Together they decided that walking around the apartment complex might be fun and a good way to meet their neighbors. For weeks the clients walked almost every day, but at different times of the day and not on days when the weather was too cold or snowy.

Later, the case manager's supervisor, a person with little human service experience, was upset that the case manager had not "scheduled the exercise." The supervisor felt that the clients should not be allowed to just say they would walk every day, "because they won't do it. We need to put that in their daily schedule, let's see, at 11:00 to 11:30 every morning. That way we can check on them and be sure they are doing it."

"What if one day they skip or feel like going at 3:00 in the afternoon?" the case manager asked. "My point exactly," replied the supervisor. "This way they have no choice, and they'll meet their goals. We'll look good for seeing that the clients' goals are implemented consistently. Set up a daily schedule for them and see that this goes in at 11:00."

A person beginning to live a normal life in a community who has decided to take up an exercise program, and who has demonstrated a commitment to that program, does not need a professional to oversee the timing of it. This is an example of inappropriate control.

A very grave example of discouragement occurred in a partial hospitalization program. Kitty, a client in the program, suffered from severe schizophrenia and depression. Often she was immobilized with sieges of despair and delusions, with voices of many others talking to her in what she called a "confused conversation." Kitty described herself as afraid and appeared to the staff as dependent. The staff surmised that because Kitty had a master's degree, obtained before her first episode of depression, she was really capable of more independence. They developed a series of goals for her to follow, such as riding the bus, shopping at the mall, and handling arrangements for her insurance and transportation. The final step on the list of goals was for Kitty to prepare her tax returns because her degree had been in business.

From the start, Kitty had problems managing the goals. Feeling extremely depressed and occasionally hearing voices, Kitty found it alarming to be on her own in the city. In group sessions, Wayne, the group leader, held her up to ridicule. He encouraged the other clients to scold Kitty and accused her of refusing to help herself more. Kitty asked to reexamine her list of goals, but this was greeted with a refusal on the part of the staff and further insistence that she "get out there and try harder." Finally, Kitty decided to withdraw from the program. When she told the staff, they told her that unless she cooperated with the program set forth for her, she would not be allowed to come to the clinic for her prescriptions. These prescriptions, partially underwritten with public funds, were important in sustaining Kitty's connection to reality.

Kitty finally called a friend of hers, a psychologist who worked in the state mental health system. When the behavior of the staff came to light, the staff was reprimanded, and Kitty was able to obtain her prescriptions and counseling services elsewhere.

We need to examine what went on in Kitty's case. First, a client has the right to ask that goals be reexamined. It may indeed be that the goals are not truly in line with the capabilities of the client, and the setting of goals should always be a collaborative effort. Second, a client can always withdraw from service if the client determines the service is no longer useful or, as in Kitty's case, actually harmful. The client has a clear right to determine what is good for and helpful to her, and what is not. Finally, the use of medication to coerce the client into doing what the staff has determined she will do is highly unethical.

All these factors combined in Kitty's case to create a discouraging atmosphere. On top of that was the denigration by the staff, particularly Wayne, who pointed out Kitty's deficiencies and ascribed manipulative motives to her failures without ever really working with her to plan goals at which she could succeed. It would have been hard for a person like Kitty to attempt any goals in such a negative situation.

Another example of discouragement happened in a transitional living arrangement for the mentally ill. This transitional living situation was one of many living arrangements with varying degrees of independence that helped clients move from hospitalization to independent living. Mario, who had persistent schizophrenia, had been placed in the last step toward independence, a small house with four other clients. At the time, Mario was on new medication, begun while he was in the hospital, that increased his ability to function considerably. While in the program, he adjusted well to the medication, obtained a job, and began to look for a new apartment. This was all part of the plan. Finding a new apartment was difficult, however; while extremely ill many years before, he had presented problems for one landlord after another. Now, in spite of the obvious improvement, the client had a reputation for being a problem when he became ill, and no one was willing to risk his becoming a tenant. In the meantime, Mario collected furniture for a new apartment and continued to look at ads in the paper. Staff in the program assisted.

The time for Mario to remain in transitional housing expired. At that point, the case manager called the assistant director of the agency. This worker was belligerent. He said the client was obviously "high functioning" and was therefore "stalling" in finding a place and moving on. He stated he knew nothing about the client's past history with landlords and was not interested. He reminded the program director that case management was responsible for the bills for the transitional housing, and they were going to stop paying for this service for this "manipulative" client. "He obviously has no intention of moving and thinks we're all too dumb to see it. Let him know he has seven days to be out."

Without asking for a meeting, without sitting down with the client or with the staff in the program, without looking at what might be going on and how the agencies could work together to facilitate a positive transfer to conventional housing, the assistant agency director wrote a peremptory and hostile eviction letter, making it clear that Mario had seven days to find housing or he would be "evicted," and stating that Mario's case manager supported this decision and thus Mario should make no appeal to his case manager for help. Coming home from work, the client found the letter on his pillow and immediately deteriorated. Staff at his transitional living program had not been informed of the letter and discovered the client in a frenzy in his room, throwing things into garbage bags and crying. Sometime later he left the house.

When he did not return that evening, the staff became alarmed. They found and read the letter left in his room, but were unsure what to do. They notified the evening supervisor of another program, and together they were able to track Mario down in another state where he had gone to be with his sister. Much later, case managers and others began to put notes in the record that indicated work had been done to find housing with this client. The notes were written to appear as if they had long preceded his leaving the program. The notes were back-dated, a highly unethical practice. These notes contained indications that the client had been uncooperative during this time, something the transitional housing staff firmly denied.

In time, the staff in transitional housing were able to let the executive director of the agency know of the true nature of the incident. The executive director had been told this was a smooth leave-taking by the client. Reprimands followed, but these in no way made up for the damage done to the client's sense of self-esteem and confidence or the ground lost in this client's move toward independence.

DEMONSTRATING WARMTH, GENUINENESS, AND EMPATHY

Instructions: There are five grade levels of responses you might give to someone who has come to you needing assistance in sorting out their problems and feelings.

Grade A: These responses are the most useful in establishing rapport and encouraging a continued dialog.

1. Centers on the client entirely
2. Stays on the topic (responds to the client's feelings or the content of what the client has said)
3. Addresses what is most important at that moment to the client
4. Is respectful (indicates the client is an equal; indicates the client is a person worthy of being understood)
5. Invites collaboration
6. Shows confidence in the client

Grade B: These responses are helpful but could be better.

1. Is somewhat confident of the client's abilities
2. Minimal invitation to collaborate
3. May briefly stray off the topic
4. Is just a little superior

Grade C: These responses are usually made by someone who means well, but the responses are not especially helpful.

1. Is pleasant, but superior
2. Overly helpful without collaborating
3. Introduces new topics that seem to the worker to be more relevant
4. Misses the feelings
5. Does not address the content

Grade D: These responses are not useful in establishing rapport and do not encourage a further exchange.

1. Takes over with solutions
2. Spends little time listening; is abrupt
3. Moralizes and preaches
4. Ignores the client's assets and strengths
5. Shows minimal interest
6. Does not indicate respect for the client

Grade F: These responses are mean-spirited and damage the relationship irreparably.

1. Uses denigrating labels and descriptions of the client or the client's actions
2. Shows no interest in the client
3. Denigrates feelings of the client
4. Denigrates the content of what the client has said
5. Intimidates, humiliates, or threatens the client (berates and scolds the client)
6. Leaves the topic for one entirely unrelated

There follow some vignettes that demonstrate the various grade levels of responses. Look at each of these responses, and assign a grade to each one. Next, using the preceding material on grade levels, tell specifically why you think the response should receive the grade you assigned to it.

Warmth, Genuineness, and Empathy

Vignette #1

A man has come to your agency for help after he was brutally beaten and robbed. The worker asks the man to tell her what happened. He describes the night the mugging took place, but as he approaches the actual incident, he finds it more and more difficult to talk.

The first worker's response: "This is really difficult for you. Would you like to wait a minute?"

The second worker's response: "Now, this is all over, Mr. Brown. It happened days ago. You need to be thinking about moving on and getting on with your life."

The third worker's response: "This must have been awful for you! Excuse me a minute." Turning to the secretary in another room, "I heard the phone ring, Sue. Was that the attorney calling? Tell him that we need that file before we can do anything for his client." Turning back to the client, "Now, where were we, Mr. Brown?"

The fourth worker's response: "I'm really wondering if you can handle this! I'm going to call mental health and set up an appointment for you. Why, you're a wreck!"

The fifth worker's response: "I can see it's difficult for you to talk about this. I'd like to work with you to see if we can find some ways to help you with some of these painful feelings. I have some thoughts that I think might help, and I'm sure you do too."

Warmth, Genuineness, and Empathy

Vignette #2

A young woman has entered the shelter after she and her boyfriend, with whom she was living, had a fight. She is badly beaten. She seemed to want to talk and has remained in the office after the worker did the admitting forms. She is rather quiet, however, and does not volunteer much information.

The first worker's response: "Yeah, another case of the violent boyfriend. Here we go again. You'd think you women would stop seeing these guys before it gets to this." F

The second worker's response: "Did he ever beat you before? I was just wondering because I'd think that if you'd been through this before, you would have left before now." F

The third worker's response: "It sounds like you've had a rough evening. Do you own your own home? No, I was just wondering if you own your own home. It says here you live in a house, and not an apartment." C

The fourth worker's response: "I was in your shoes once. Believe me, it was a long and difficult battle to get out of that mess. But I did. I just decided that it wasn't worth living like that—life's too short, and I got out!" C

The fifth worker's response: "What you need is a good lawyer. You tell the staff that comes on in the morning that you want to talk to a lawyer. You can't just sit around and take this stuff!" F

Warmth, Genuineness, and Empathy

Vignette #3

A rape victim sees the volunteer in the emergency room. As the volunteer talks to her, she learns that the victim is afraid to go home. The man who raped her is an acquaintance in the neighborhood, and she is afraid that now that he knows she called the police, he will come after her.

A   *The first volunteer's response:* "It sounds like you could use a place to stay tonight. Would you like me to try to see if I can set something up for you?"

B   *The second volunteer's response:* "You're really afraid of this guy! Well, you're not going home tonight."

C   *The third volunteer's response:* "I can see that you can't handle this! I'll make all the arrangements. Don't worry about a thing. I'll get you a place to stay, and I'll set you up with an appointment at mental health."

The fourth volunteer's response: "How would you like to handle this tonight?"

The fifth volunteer's response: "Let's get off this gruesome topic. I mean, I know it's important to you right now, but it'll do you good to talk about something else. Tell me about your job."

The sixth volunteer's response: "You must be feeling so afraid of him."

Warmth, Genuineness, and Empathy

Vignette #4

A man talks to the intake worker in a case management unit about his recent separation from his wife. He comments that he could have been a better husband, that he was stingy and went for days not speaking to his wife if he was annoyed. He thought he was making her see how he felt about things, but she left him, saying he was "uncommunicative." He seems depressed and bewildered by his wife's departure.

The first worker's response: "Sounds to me like you could use some communication workshops."

The second worker's response: "Did you have to go that long without speaking? I mean, what could she have done that made you that mad?"

The third worker's response: "You really thought you were getting through to her, so it must be hard to see her leave this way."

The fourth worker's response: "Only a fool would think what you were doing was 'communication'! Of course she didn't hear you, fellow!"

The fifth worker's response: "See, I think you should have tried marriage counseling *before* things got this bad. Not now, after she's already gone."

The sixth worker's response: "Well, if you had a better understanding of the way women think, you could have avoided this whole thing."

RECOGNIZING THE DIFFERENCE: ENCOURAGEMENT OR DISCOURAGEMENT

Instructions: Following are two vignettes. Decide what you would *do* or *say* to encourage the person. Decide what you could *do* or *say* if you were discouraging the person. Actually picture an encouraging or discouraging person. Remember your actions may be as important as your words. Write in your answers to share with the group.

1. A woman, the mother of two children, has been without a home for a number of months. She tells you she really wants a permanent place to stay. You know there are very few places she can go. You also know she has some talents and interests—strengths that might be to her benefit.

   You encourage her by _____

   _____

   _____

   _____

   _____

   You discourage her by _____

   _____

   _____

   _____

   _____

2. A man calls and says he was sexually abused as a child. It has come to haunt him recently, but he is not sure where he should turn for help.

   You encourage him by _____

   _____

   _____

   _____

   _____

   You discourage him by _____

   _____

   _____

   _____

   _____

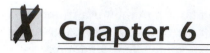 **Chapter 6**

Seeing Yourself as a Separate Person

INTRODUCTION

There is always a danger that we will see ourselves in the clients we serve. Many of us will go to work in agencies that serve clients who have been through something we went through ourselves. We may meet a client whose situation is different but that person reminds us in some way of ourselves. Perhaps the person is our age or has the same interests we do. When clients remind you of yourself and you are unable to separate your circumstances from theirs, you will become a problem to the very person you are supposed to be helping.

Sometimes we have not entirely resolved the issues in our own life. We may seek employment in an agency that deals with similar problems just to continue to heal. This is not useful. It means that the clients will be treated in the light of your own issues rather than a completely objective focus on their own personal issues.

SEEING YOURSELF AND THE CLIENT AS COMPLETELY SEPARATE INDIVIDUALS

The Client Reminds You of You

In a training for volunteers at a rape crisis center, a young woman who was training to answer the hot line had been raped some years before. It was still an overwhelming event to her, the defining event in her life. She was not ready to begin to work with others who had been raped while her own emotions were so raw and intense.

During the training, she would fret and stew over the information being given. She would constantly remind the trainer that this would not have worked for her in her situation because she was too upset or too badly injured. She pointed out her own circumstances and told the trainer to focus more on this type of situation. She gave details of her rape at every opportunity. She used her own situation to illustrate the trainer's points.

It was apparent that she was not ready to serve as a volunteer who would have direct contact with the clients. Why? Because she had not yet recovered from the trauma of her own rape, and it seemed likely that she would impose the circumstances and emotions of her own rape onto those of the caller. In that case, the caller's real problems might get lost as the worker went off on her own concerns and feelings. She was asked to take other work in the agency, and ultimately she left because she never was able to separate her rape from those of the clients the agency served.

Sometimes the client is trying to do something we accomplished ourselves long ago. The client reminds us of a time when we were vulnerable and uncertain. For example, in a welfare agency, a young woman, recently off welfare and now working as an income maintenance clerk, was helping a welfare client make plans to get off welfare. The worker was somewhat irritated by the client's concerns about day care and transportation. "Look," she finally blurted out, "If I can work every day, you certainly can!"

In another situation, the similarities were too much for Yolanda, a human service worker, to tolerate. She went to great lengths to find differences between herself and the client. Working at a shelter for victims of domestic violence, Yolanda, who worked the evening shift, had trouble understanding how a shelter resident could have "gotten into a mess like this." The resident had the same education as Yolanda, and the resident's husband and Yolanda's husband held very similar positions with accounting firms. Yolanda and the client were the same age, and both had two children, ages 6 and 8. The client came from a middle-class suburb not far from where Yolanda lived. Yolanda was unable to believe things were as bad as the client described them; she said that if they "are half as bad as you say, I can't see how you let this happen!" In spite of hours of training in which Yolanda learned that domestic violence can happen in any socioeconomic group, she nevertheless confided to coworkers that she thought the resident was "stupid" to have married such a person in the first place and that the resident and her family must be "pretty lower class, because people like that don't generally live in my neighborhood."

In this case, Yolanda found it frightening that these problems could exist so close to her home among people so much like her. She found it necessary to protect herself from this reality by looking for extenuating circumstances that would indicate a clear separation between herself and the client. In working on this, she missed being really helpful to the client.

The Client Reflects on You

There are other reasons that a person might not separate from the client. As a case manager, you might become extremely involved in clients' problems and solutions because you will feel more important or more competent if the clients solve their problems successfully. It may impress others if all your clients do well, and so you may cross the boundary and force clients to use your solutions. Sometimes, in spite of what we know about how good it is for clients to learn from their errors or struggles, we mistakenly believe that we must have all happy successful clients to make ourselves look good. Again, we are focusing inappropriately on ourselves, and not on the client's needs.

POOR ASSUMPTIONS

Some boundaries are artificial and foolish. There is a very human tendency to make two extremely unfortunate assumptions about people. These two assumptions are:

1. Assuming that people who look like you will think and act like you

2. Assuming that people who do not look like you are not like you at all, but are very different

Be very careful what you assume to be true of another person. A person may be of another race or culture and yet share many of your values and circumstances. A person who looks like you may be quite different in tastes, opinions, and way of life. If you assume a person of another race or religion has certain stereotypical characteristics, you will be dealing with a stereotype, and not with a real person.

The tendency to view people in light of these assumptions means that you no longer see individual differences. All Catholics, African Americans, Jews, Quakers, or Indians are not alike. If you view a person based on your stereotype of the group from which he or she comes, you have failed to discern individual differences. Failure to perceive individual differences is a failure to accurately perceive reality. Aside from this being a fundamental characteristic of mental illness, it is impossible to give excellent service to a client you have seen only as a stereotype.

SUMMARY

It is important, for all the reasons stated in this chapter, to view each client as an individual person and to work very hard to understand what makes your client a unique human being. Using the individualistic and collectivistic models can help you to understand your client better, but then you need to recognize, through careful listening and observation, what it is that makes your client a unique person. Professionals work against relying on stereotypes, assumptions about groups of people, or personal feelings about certain problems and their solutions that will color the work they do with others. Take the time to see and hear the characteristics, circumstances, and interests of this person who is trusting you to assist in some way. Focus on what makes your clients separate individuals. Give your service based on what you know about your client as a unique person.

When workers begin to identify too strongly with their clients or find that they must assume some of the clients' responsibilities, they have blurred the boundary that makes both the worker and the client separate individuals. They cross that boundary in unproductive ways every time they assume that the client will be like them or will react as they would. These workers breach the boundary each time they handle a client's problem without collaborating with the client.

BLURRED BOUNDARIES

Instructions: Following are some situations in which the boundary between the worker and the consumer has become blurred. Identify what went wrong and what needs to happen to correct the situation. Use the space provided to make notes.

1. Alice is very upset because she gave the consumer some names of people she thought might be helpful in solving the consumer's problem. Today she met one of those people at the local deli where she eats lunch and learned that the consumer has never been in touch. She has been trying to call the consumer all afternoon to find out what happened.

2. Bill is feeling very proud of himself. He talked to a consumer who had very complicated problems this afternoon. He put everything down on paper, while the consumer sat by his desk and drank a soft drink. Then Bill decided on the best way to handle the situation. The consumer finished his drink, did just what Bill suggested, and reported that everything is fine now.

3. Mary Lou was once in a very abusive relationship. She was able, through much counseling and grit, to become assertive enough in her own behalf to get out of the situation. Today she is happily married and the mother of two lovely children. The consumer she is talking to is in just such an abusive situation and seems hesitant about leaving. Mary Lou wants her to leave and, using examples from her own life, assures the client she is certain that the client will have just as happy a life as Mary Lou has had if she will leave. Mary Lou remembers vividly how she felt when she was in the client's shoes and tries to make the client see how much better she will be if she leaves now.

4. Gloria was raped by her stepbrother when she was 16 and he was 23. It was a very difficult situation; law enforcement officials were called, and eventually the situation broke up her family. She is currently working at a rape crisis center and is talking to a consumer whose situation reminds her of her own. She says things like: "Oh, that wouldn't have done a thing for me!" or "I was too far gone to be able to handle it that way."

5. Carlos is working with a woman whose father is the president of a large bank in another city. She has become depressed and needs a referral to a therapist and perhaps a psychiatric assessment. Carlos is reluctant to talk to her about needing "psychiatric help" because he assumes this would be offensive to someone as "upper class" as she is. Instead he suggests she "talk to someone for a little bit." Later, with the woman's permission, her father contacts Carlos and asks if there is anything he can do. Carlos is careful not to suggest short-term psychiatric hospitalization even though the father is offering to pay "for whatever she needs." Carlos is sure such a hospitalization would alienate the father.

6. Candy is working in a shelter for homeless men. She is new at her job and she enjoys what she is doing. The fact that many of the men are in poor physical health and are unwashed is of great concern to her, and she works hard to meet the basic needs of the residents, such as food and clothing and a warm place to sleep. When

she is working, she never asks the men to help with chores around the shelter even though that is a condition for their staying there. She thinks of her work as very loving and giving, and she sees the clients as hapless and uneducated. Therefore, when Paul comes to her and tells her he graduated from high school and finished a year of college before he got hooked on drugs, she is not certain she believes him. When he asks for help returning to college, she resists giving him information and support for attending the local community college.

7. Tom goes to the home of an elderly Mennonite woman to do an assessment. She is bedridden and will need in-home services. The family is there as well, and Tom talks to all of them. The conversation turns to a neighbor who is in the military and who has been harassing the family recently about a fence the family put up for the woman's dog about four years ago. Because the family is Mennonite, the family members express pacifist views about the military and express their perception that military people are "argumentative—they like to fight." Tom heartily agrees with them, pretending he never served four years in the army. Later, at the office, he confides to a colleague, "People like that can never really stick up for themselves. They are pretty much out-to-lunch."

Chapter 7

Clarifying Who Owns the Problem

INTRODUCTION

Before you ever open your mouth, before you ever say a word to your client, you must be able to discern accurately who owns this problem. Who owns the problem?

> *It is the person whose **needs** are not being met.*

It is not the person who is being rude and uncooperative. It is not the person who is ruining a party. Nor is it the person who is singing off-key and ruining the song fest. It is the person whose needs are not being met.

There are three good reasons why you should know who owns the problem:

1. *You know who is responsible for solving the problem.* If you know who owns the problem, you then know who is ultimately responsible for solving it. If you know who is responsible for finding a solution, you will not assume the entire responsibility is yours. In other words, you will not accept responsibility for problems that are not yours. When you take over and try to solve other people's problems, or tell them how they should resolve their problems, you could be seen as meddling in their affairs or pushy.

2. *Meddling is disrespectful.* This sort of meddling is disrespectful, even when you intend it to be helpful. It says clearly that you have your doubts about the other person's ability to figure this out and handle it on his or her own. You are not sure other people have the sense and insight to know what is best for them.

3. *The client loses opportunities to grow.* Furthermore, when you take over with solutions, you interfere with what might be a very meaningful experience for the client. This person may grow from wrestling with this issue. It may be the opportunity needed to gain insight, learn a new skill, or try something that until now has been too frightening. If you are attempting to take over with your own solutions and ideas, your client misses this valuable opportunity. Clients can never say, "I did this myself!" Instead, they will have to say, "My case manager did this for me."

Kentaro, working in a sheltered workshop, was learning a new job on the assembly line. He seemed to quickly pick up his responsibilities, but was having trouble keeping up.

Over and over the worker monitored his progress and gave him tips for improving his speed. The worker stood behind him and grabbed the pieces Kentaro missed. Finally, the worker was called away to the phone. When he came back, he discovered that Kentaro was sitting so that he faced the assembly line from a different position. Now, with better visibility, the client was catching each piece that came toward him and making the necessary adjustments. The worker later said he felt foolish for standing over Kentaro all morning when it turned out the client knew all along how to solve the problem.

Agnes wanted to have a better relationship with her mother. She confided this to her worker one day and the worker set about helping her solve the problem. While the worker spoke to her about poor communication, mother and daughter relationships, and family therapy, Agnes decided to buy a pretty card and send it to her mother. In the card she told her mother how much their relationship meant to her and how much she wanted them to be friends. She enclosed a little lace handkerchief, and sent the package off to her mother. Soon her mother called Agnes, and they began to talk. Agnes, who knew herself and who had lived a good portion of her life with her mother, understood how best to solve her problem. Listening and helping Agnes talk about the relationship with her mother might have been a better course of action for the worker.

Keep your clients in a position of authority over their lives to the greatest extent possible. Remind clients of how much of the resolution of the problem is their own doing. Make sure your clients have the opportunity to feel pride in their part in solving the problem. Let clients see they can help themselves more (even if it is only a little bit more) the next time, rather than turning to their case manager to solve the problem for them.

IF THE CLIENT OWNS THE PROBLEM

Let us suppose your client comes in and tells you she cannot stand living with her mother anymore. Her mother is verbally abusive and rejecting. Your client is unhappy. Obviously this client's need for a pleasant home environment and her need to be appreciated by her mother are not being met. The client owns the problem.

Does the mother own the problem? She does not appear to. This method of communicating with her daughter seems to work for her. She shows no discomfort or guilt about any pain she might be causing her daughter. It seems to meet some need of hers to communicate in this way. The mother does not have a problem in this situation, as her needs appear to be met.

There are several important ways you can respond to your client's situation. First, listen. Then, rather than providing a solution, be a resource to your client. Give her options. Tell her about services with which you are familiar that might be helpful to her, but leave the final decisions up to her. In this way, you make sure that the client retains a position of power in her own life, and you act collaboratively.

Now let us change the story a little. Your client brings in the same problem, but she has a moderate developmental disability. The problem still belongs to her, but now you make a conscious decision to get a bit more involved and a conscious decision about the extent to which you will get involved. Just because someone tells you about a problem does not mean you must solve it. As a case manager or worker, there will be times when it is important to give more help than others. A wise worker will know how much to help and when to stand back. These are strategic decisions.

Deciding how much to become involved is important when the client has a problem with you. For example, a client may come into your office and tell you that he cannot stand the way you sit in front of a cluttered desk and talk to him. He claims he feels disorganized by your clutter and wants you to have your desk cleaned off when he comes in. In this case, you may make a conscious decision to let him own this problem because there is no way to clean the desk off in the middle of a busy day when you know he is coming in. You would thank him for his comments, give a word or two about why that might not be feasible, and tell him you will continue to keep your office as it is for now.

On the other hand, the client may be upset with you because you are always late. His need for punctuality is obvious; your being on time means to him that you are expecting him and that you value his time. His need is not being met. In this situation, you might decide to help own the problem. You recognize that you have been somewhat disrespectful. You can justify it with your busy schedule, but you also can do better. So you acknowledge the problem, thank him for his comments, and offer to be more punctual. You have made a conscious decision to become involved in the solution to his problem.

It Is Not Uncaring

Sometimes we feel guilty about not doing more. Sometimes others tell us that we should be doing more. After all, we are the person's case manager. Why are we not extending ourselves further? Sometimes the clients themselves are the ones to accuse us of not caring or of being indifferent. Knowing who owns the problem and allowing that person to resolve it is not an uncaring action. In fact, you would never refuse to help a client simply because you determined the problem belonged to the client.

When you allow clients to work on their issues and problems, however, you respect the clients' right to privacy and self-determination. In addition, you give these people an important opportunity to grow and work in their own behalf. Solving one's problems effectively is part of emotional health and maturity. To the extent to which clients are able, we want to encourage them to do as much for themselves as they can.

It Is a Strategic Decision

The extent to which you become involved in helping clients solve problems that belong to them is a strategic decision. This is another difference between the professional approach to relationships and a friendship you might have away from work. In the professional relationship with your client, you want to decide strategically how much help to give and the extent to which you will step in. The decision is based on your knowledge about the client and about how this opportunity can be used to help your client grow.

A woman who is blind might need more help negotiating the transportation system than one who is depressed. A person who is illiterate and from a rural village might need more help working with the Social Security office in the city than an urban lawyer. A child might require more support to carry out a personal decision than an adult.

The strategy lies in knowing your client's strengths and limitations and tailoring your involvement to those factors specifically. In this way, you avoid taking over simply because that is the easiest thing to do or because you see all clients as helpless. Your involvement is just to the point the client needs help or ideas and no further.

You will find that in certain cases where the client owns the problem, you will take over and resolve it almost entirely alone. Suppose you are working with a 17-year-old single girl, disowned by her family because of her pregnancy. She has just delivered her first child for whom she has made adoption arrangements. The child, however, is severely disabled and retarded. The doctors feel the care required by the child can never be undertaken by a 17-year-old girl living alone, and the prospective parents have now withdrawn their bid to adopt the infant. In this case, you work out arrangements for the care of the infant, solving the infant's need for a safe, medically appropriate environment and solving the mother's problem of what to do with a handicapped child she believed would be going to the home of another couple. In this case, you would consult with the mother throughout the process, possibly even taking her to see the facility where her baby will receive care, but you would handle most of the actual arrangements.

If you did not understand the concept of who owns the problem, you might be tempted to ignore the mother in the process of solving this problem. If the mother were older, married, having her second child with this handicap, and supported by her family (or in any number of different circumstances), your response and the extent of your involvement would change as well.

When the client owns the problem, *carefully* decide the extent to which you will be involved. Test your hypothesis about how much the client can handle alone. Be ready to take on more responsibility or give more responsibility to the client as you move toward a solution. Watch your involvement to be sure you are not obstructing the client's opportunities to grow or to exercise self-determination and independence.

Be a Resource and a Collaborator

You will have at your fingertips information that can help the client solve a problem. You may have the names of agencies, phone numbers, contact people, and addresses for services. You will also be familiar with policies in various agencies and within large social service systems, such as child welfare and mental health. You will often be more familiar with the law as it pertains to the client's situation. This makes you a valuable resource to a person attempting to arrive at a solution.

Bring the information and facts to the client and then collaborate with the client on the solution. It is the client who is most aware of which solution will work and which ones will be impractical. Together, with your knowledge of the system and your clients' knowledge of their personal lives and circumstances, you will be able to construct a useful approach to clients' problems.

IF YOU OWN THE PROBLEM

If you are having a problem, that is, your needs are not being met, you will understand that the resolution of the problem is ultimately your responsibility. This applies to personal problems, and it also applies to problems you might have in the course of your relationship with your clients. What if it is the client who is always late? Whenever he is late, you find you are behind for the rest of the day. This is not the client's problem. He may find it perfectly acceptable to get to your office at approximately the time he is scheduled to see you. He may have scheduling problems or punctuality problems; but in this case, he does not own this problem. You do. Your need to stay on schedule and see everyone you are scheduled to see before 5:00 P.M. is not being met. Therefore, you are the one who is responsible for bringing it up. Do not expect that others will guess there is a problem.

In bringing up a problem we are having with another person, we are asking for that person's assistance in resolving it. Just as you make decisions regarding how much you will become involved in resolving someone else's issue, your clients have the right to determine the extent to which they want to help you. It is conceivable that the client will see your point and make some changes. It is also possible that the client will decide that it will have to be your problem because it is preferable to be late for whatever reason or because being punctual is an inconvenience. There are ways to solve problems like this one; but for now, as the first step, you need to be clear about who owns this problem.

IF YOU BOTH OWN THE PROBLEM

Sometimes you both have a problem. He needs evening appointments, and you work only during the day. She wants to shout and yell about her situation, and you find that too unnerving to do a good interview. These are opportunities to negotiate. You, as the worker, have to be able to sort out in your own mind who owns what problem, and you must be able to initiate some negotiation around these issues.

When you both own the problem, you should not view it as a win/lose situation. If the client sees it that way, you need to point out other ways of looking at the situation. Perhaps he can see another worker who does work at night, or perhaps he can come in during the day sometimes and you can stay late sometimes. Maybe she can yell with less intensity, and you can overlook the rest of it. It might work to transfer the client to another worker, one whose schedule is better suited or one who can better tolerate the yelling.

There are many ways to negotiate a solution. When you work on a solution collaboratively with the client, you provide the client with an important experience in problem solving. As the worker, you invite the client to join you in this effort.

Margaret had been ill with schizophrenia for a very long time. Rejected by most of the community and most of her strictly religious family, she found solace and support among the workers in the mental health system. In the course of her illness, she had been hospitalized and knew the staff at the hospital well. She had encountered the various members of the crisis team and knew them, too. She had a case manager whom she found supportive.

Margaret found countless reasons to call workers within the system. Night and day she called with tiny questions, not so much because she could not resolve the problems herself, but because she found contact with these supportive people comforting. Sometimes she would call to ask what time she should go to bed. She might call to ask if she should eat one frozen dinner rather than another. Should she go out for a walk tonight or not? Should she buy a new pair of shoes or not?

Margaret's incessant calling began to create a problem for already busy workers. They grew exasperated. Margaret had a need to feel their support, and the workers had a need to get things done with other clients. Finally, a solution was worked out with Margaret and all the workers in the system who regularly received calls from her. It was decided that a man on the crisis team who shared her religion and genuinely liked her would be the person she would call. When he was off-duty, a backup person was designated. Margaret was then allowed only one call a day. She was to save all her questions for that one call. Everyone agreed to this plan.

Although Margaret tested the plan many times initially, everyone stuck to the agreement. Eventually Margaret began to make the calls more meaningful, asking for help with real problems. Undoubtedly, this one call a day helped to sustain her and helped her to live more comfortably in the community rather than in an institution. It also allowed the workers to focus on other clients.

In this situation, both the workers and the client had a problem, and their needs conflicted. By collaborating on a solution, rapport was not lost, and both the workers and the client gained valuable experience.

WHO OWNS THE PROBLEM?

Instructions: In the following situations, identify who owns the problem. As you study each case, decide whether it is you, as the worker, who owns the problem; or whether the consumer/client, and perhaps his or her family, owns the problem; or whether both you and the client own the problem at the same time.

1. A woman you have placed in temporary housing is angered by the loud music of her neighbors. She appeals to you to do something about it. Who owns the problem?

   WOMAN

2. You work at a victim/witness resource center where you assist the victims of crime to handle the emotional and technical ramifications of the crime before they go to court. The husband of a victim, a woman who was car-jacked by a teenager one night, takes you aside and asks you to persuade his wife to drop the charges. He tells you confidentially that it would be better for his wife if "she didn't have to go through this." Who owns the problem? *husband*

3. The mother of a rape victim, with whom you have been working, calls and says that ever since the rape, her daughter has been crying and unable to eat or sleep. She tells you it is urgent that she know exactly what happened to her daughter, but that her daughter refuses to talk about it. She asks if you can tell her what happened. Who owns the problem? *mother*

4. You are talking to the victim of a violent crime in the emergency room. Her boyfriend barges in and demands to know "what's going on." Who owns the problem? *boyfriend*

5. You have placed a woman in temporary housing after she left her home following severe abuse by her husband. The husband calls demanding to know where she is and tells you he will get his lawyer and sue you if you do not tell him. Who owns the problem? *husband*

6. You are working with a support group. One of the participants tells you on the side that another participant is monopolizing the group's time with frivolous details and asks if you will do something. Who owns the problem?

   the first participants

7. You accompany a victim to court. The defendant's family meets you in the hall and tells you that you are responsible and the agency is responsible for their son being on trial. They claim the whole thing is trumped up, and they think you should persuade the consumer to drop the charges. Who owns the problem?

8. You are arranging for housing for a woman who is in a homeless shelter. Her parents come to see you and ask you to see that she also goes to therapy. They tell you she has never "seemed right," and they ask you to give them your opinion of her mental status. Who owns the problem?

9. You have developed a goal plan for a child. The parents agree with the plan, which involves summer camp and other recreational activities over the summer, all with a therapeutic program. The teacher calls to tell you that this child can hardly benefit from school and that sending him to camp is a waste of the taxpayer's money. What he needs, she tells you, is therapy. Who owns the problem?

MAKING THE STRATEGIC DECISION

Instructions: Following is a basic situation, with a list of scenarios in which the circumstances surrounding the situation are different. Decide what you would do in each case.
Situation: Hannah recently went blind due to an accident with chemicals at the company where she worked. She is asking for a service plan that will help her regain some independence.

1. Hannah is a Ph.D. chemist with the corporation where the accident occurred. She has received a huge settlement from the corporation's insurance company. The corporation has said she can come back to work if she can be retrained in some way, possibly with computers. Hannah has a supportive husband and many close friends. How do you help? *Get training to read braiel*
   Get a eye-seeing dog
   Get computer training
   Get transportation to work and back
   Have home redesigned for a blind person

2. Hannah was a custodian at the small chemical engineering company where the accident occurred. The company had no insurance, and it has no interest in hiring her back for any reason. The company did give her $5,000 at the time of the accident, and the hospitalization plan and workers' compensation helped pay the initial medical bills. Hannah lives alone and has few friends. How do you help?

3. Hannah is mentally retarded and worked as a custodian at the company where the accident occurred. The company gave her $5,000 at the time of the accident, and the hospitalization plan and workers' compensation helped pay the initial medical bills. Hannah lives with her parents, who are very supportive, and she has two older siblings who also give support. The family has been working with Hannah to help her decide what to do next, and they have found a place where she can answer the phone and give standard information. This company is delighted to

have a real person to do this, as the answering machine option seemed too impersonal. Hannah will need some training. How do you help?

4. Hannah is a student working on a chemical engineering degree and wants to remain in school. Her family is supportive of this, but they live in another state. Hannah's roommates seem hesitant about her returning to live with them in their downtown apartment now that she is blind. How do you help?

5. Hannah is a student working on a chemical engineering degree and wants to remain in school. Her family is supportive of this, but they live in another state. Hannah wants a seeing eye dog, has a landlady who is afraid of dogs but who might accept one, and needs to learn how to negotiate the town and the campus as a person who is blind. She has numerous supportive friends. How do you help?

6. Hannah is a student working on a chemical engineering degree and wants to remain in school. Her family is supportive of this, but they live in another state. Hannah was using this job to pay for her education. Now Vocational Rehabilitation will help, but Hannah must fill out countless forms. Hannah is depressed and frightened by her blindness and spends days at home alone. How do you help?

7. Hannah had a fairly ordinary chemical technician's job at the company when the accident occurred. She and four other people were blinded by the accident. Hannah has told you she wanted a lawyer while she was still in the hospital, and she also felt the group should meet regularly to talk about the accident and their anger. The others have agreed. Hannah tells you of the state office of services to the blind and wants help connecting to that office. How do you help?

Section 3

Effective Communication

 **Chapter 8**

Identifying Good Responses and Poor Responses

INTRODUCTION

Following are some poor responses, poor in the sense that they tend to block communication and prevent the worker from hearing what is really important. These poor responses are followed by good or constructive responses that encourage another person to talk and feel comfortable doing so. Like learning to drive a stick-shift car, we learn step by step the ways to structure responses that will facilitate good communication.

It is not enough, however, for you to look at these responses and do the exercises. You will build a skill in identifying which responses are inadequate and which actually enhance the communication, but this will not teach you to automatically use good responses rather than poor ones. This chapter is a start, but there is no substitute for practice. In later chapters, you will learn to use these constructive responses with more understanding of how they promote rapport and clarity. Even this understanding does not help you to make an automatic response that is therapeutic if you have not practiced.

What we will do in this chapter is to clarify which responses are most likely to promote rapport and which are most likely to promote withdrawal and defensiveness.

THE TWELVE ROADBLOCKS TO COMMUNICATION*

Dr. Thomas Gordon's (1970) *Parent Effectiveness Training* outlined twelve specific ways we often block good communication, setting up barriers to real understanding and dialogue. We have adapted these roadblocks to fit the kind of poor communication that sometimes happens in the human service setting. These responses are not helpful in talking with other people. They serve to obstruct rapport and block any constructive resolution of the problem, and they often serve to make our clients feel inferior or demeaned.

What is important to notice in so many of the responses that follow is the implied superiority of the worker. Individuals who come seeking assistance already feel unsure of themselves and uncertain about what to do. Workers who come across as all-knowing or judgmental are not helpful. It is the attitude of superiority, a sense that the worker is talking down, that is harmful to the relationship and makes real communication and rapport difficult.

---

*Adapted with permission of Random House, Inc.

In addition, notice how hard it would be for a client to trust and be open after hearing these responses. What does one say to a person who appears to be convinced that he or she is superior? Most clients stop talking or resort to pleasantries after that. Real communication is blocked. Why continue being open if the responses are not empathic? We can look now at these negative responses.

Ordering, Directing, Commanding

The first roadblock involves giving the person an order or command. Here the workers take charge without including the client. The assumption these workers make is that they have all the correct answers or all the best solutions and ideas. There is no dialogue or collaboration.

> "I don't care what anyone tells you! You have to go see that lawyer!"
> "Go right back over to the court house and get those forms!"
> "Leave your house and find another one."
> "Look, just go over and apply for the job."

Warning, Admonishing, Threatening

Warning of consequences if the person does something is the second roadblock. In this series of responses, the workers are again operating from a superior position without consulting their clients. They do not want the clients to follow certain lines of action. Rather than discuss their clients' inclination to act in a particular way, they are warning the client instead.

> "If you take that suggestion, you'll be sorry!"
> "You'd better not do that if you know what's good for you!"
> "I can tell you something like that won't work!"
> "I can tell you right now you're headed in the wrong direction."

Exhorting, Moralizing, Preaching

The third roadblock is telling the person what he or she should or ought to do. Again the workers display the belief that they are superior to their clients and have all the answers. In several of the responses, we can hear workers imposing their own moral values on the client. Listen to the "shoulds" and "oughts." Workers using these two words speak as though what they are saying is a universal given, rather than a personal choice or value.

> "You should know that doing that is wrong!"
> "You shouldn't think like that."
> "You ought to see a counselor."
> "You ought to be more concerned."

Advising, Giving Solutions or Suggestions

Telling the person how to solve his or her problem is a fourth roadblock. When workers believe their clients have nothing useful to contribute to the resolution of a problem, they will make unilateral decisions without their clients' input. In some cases, the following responses indicate exasperation with the client. These workers believe that the way they see the clients' problems is the only way to see them. The clients are hapless or inadequate for not seeing them that way as well.

"Why don't you just call your husband?"
"I suggest you stop seeing this person today."
"It would seem to me that you should just stay home."
"Why don't you simply tell him to stay away?"

Lecturing, Teaching, Giving Logical Arguments

The fifth roadblock involves trying to influence with facts, arguments, and logic. In this series of responses, the workers sound as if they believe their clients are incompetent. Only the workers know the whole picture. These responses do not encourage real discussion of a person's feelings and problems.

Look at the facts about domestic violence."
"Look at it this way, the longer you put up with this, the more she gets away with it."
"Now what you need to do is call the police. That will circumvent any action on their part and free you to move to another location."
"Now look, you have two choices. You can either stay or leave. That's what you have to decide."

Judging, Criticizing, Disagreeing, Blaming

Making a negative judgment or evaluation of the person is the sixth roadblock. In the responses that follow, the workers see themselves as judges of the clients' behavior. Instead of being supportive, these workers are grading their clients' behavior. Their responses can only serve to demean a person who is grappling with problems and feels unsure.

"You aren't thinking clearly."
"You're very wrong about that."
"I couldn't disagree with you more."
"Your plan is faulty."

Praising, Agreeing

The seventh roadblock is offering a positive judgment or evaluation, or agreeing. Sometimes workers really cannot tolerate the pain a client expresses about a certain situation. The responses that follow are certainly well-meaning, but they cut off meaningful discussion and relieve the worker from having to deal with the real pain the client might want to talk about.

"Well, I happen to think you did just fine."
"You'll figure this out."
"It will all work out for the best."
"You're smart enough to know what you need to do."
"You're bright. You know what to do here."

Name-calling, Ridiculing, Shaming

Making the person feel foolish is the eighth roadblock. Perhaps the workers who give responses like the following are fed-up. They may tell themselves they have a right to express these degrading sentiments because they put up with so much from their clients. Only very untrained and unprofessional workers would ever resort to name-calling, but when it happens, these workers generally say they have endured long-term disgust or exasperation as a way of excusing what they have said.

> "You're an idiot to do that."
> "Okay, smarty, do it your way. You'll see."
> "What you're doing is totally ridiculous!"
> "You are the silliest person!"

Interpreting, Analyzing, Diagnosing

The ninth roadblock consists of telling people what their motives are, or analyzing their actions. Sometimes, in order to feel vastly superior to their clients, workers will engage in surprise revelations. It is often done to show the client that the worker knows more about the client's inner conflicts than the client does. In these responses, the workers are informing the clients of their motives and underlying intentions as if the clients lack self-awareness.

> "You're just upset because you haven't heard from the lawyer."
> "I think you really wanted to press charges but you can't admit that."
> "I don't think you really believe that about her. You're just saying it."
> "What you really mean is that you don't want to see him anymore."

Reassuring, Sympathizing, Consoling, Supporting

Trying to make people feel better or trying to talk them out of their feelings is the tenth roadblock. The following responses are offered as a way of comforting another person, but they serve to cut off real discussion of painful feelings. Telling someone you know how he feels or you understand what she is feeling is not convincing, even if you have had similar experiences. Listening to the feelings is better than cutting them off.

> "You'll feel better in the morning."
> "All new mothers go through this at one time or another."
> "Don't worry. Things will work out."
> "I know exactly how you feel!"
> "I understand how you feel."

Probing, Questioning, Interrogating

The eleventh roadblock concerns trying to find motives, reasons, and causes. Clients do not always tell their concerns in logical sequence, and there are good ways for workers to go back and fill in the gaps without sounding as though they are in a superior position. Here the workers are actually prying into motivation and intention, areas the client may not be fully aware of or ready to discuss.

> "Just when did you start to feel this way?"
> "Why do you suppose you went there that night?"

"Do men ever tell you that they feel violent toward you?"
"What were you really trying to do when you saw her?"

Withdrawing, Distracting, Humoring, Diverting

Trying to get the person to focus on something other than the problem is the twelfth roadblock. Sometimes workers are overwhelmed by what their clients have told them. They may feel helpless to make a real difference or offer substantive help. Perhaps they realize that all they can do is listen, but listening is painful and difficult. In order to save themselves from these uncomfortable feelings of inadequacy and helplessness, they resort to responses like the ones that follow.

"Just forget about it!"
"C'mon! How are things at church?"
"Let's see, you could always run over him in your car." (chuckle)
"Let's turn to other things in your life."

IDENTIFYING ROADBLOCKS

Instructions: Look at the responses that follow and decide if the worker is blocking communication or enhancing it. Check your answer after each vignette.

1. Carlos is afraid his mother is dying. He is talking to the worker in the hospital emergency room about an "attack" his mother seemed to have when she could not breathe and turned blue. She was brought to the hospital in an ambulance, and Carlos is waiting to see whether she will be all right. He is distraught. The worker says, "You certainly did the right thing to call the ambulance. Don't worry she'll be all right. We have very good doctors here."
   Enhanced   Blocked

2. Anita wants to change schools to be with her old friends. Her parents feel she is not adjusting to the move and asks the worker to talk to her about her desire to return to her former school. Anita talks about how strange the new school is and how much she misses her old friends. The worker replies, "Tell me something about your friends where you used to live."
   Enhanced   Blocked

3. Elvita has decided to leave an abusive relationship, but she feels guilty leaving her abuser's children behind. She talks about how she knows that once she leaves, she cannot have any more contact, but worries about how that will affect these children whom she has come to love and protect. The worker asks, "Just when did you start to think of these children as if they were your own?"
   Enhanced   Blocked

4. Ed suffers from chronic mental illness and has a need for medications to maintain his mental health. Recently he went to several workshops looking at how to use supplements and vitamins to maintain mental health. He wants to discuss these ideas with his worker and is clear that he is not really thinking of going off his prescriptions, but would like to consider trying these supplements in addition. The worker says, "It sounds like you really got a lot out of that workshop."
   Enhanced   Blocked

5. Shawna wants to go to college. She went to a poor rural school where most of the students did not go on to college. Her scores for the entrance exams are very poor in math, and she feels unsure that the developmental course being offered to her at the college will really help her to catch up. She seems anxious and uncertain. The worker says, "You don't seem to quite understand what developmental courses are. Look at the number of people who take them. Look at how many of those people finish school. You need to think about this a little less emotionally."
Enhanced   Blocked

6. Ada is very upset over the divorce settlement. She got the house and the children, but very meager support and only a small amount to go to school to upgrade her skills. She is crying and expresses the belief that she cannot make it. The worker replies, "Look, get another lawyer. You're going to have to just face the fact that you had a lawyer who wasn't serving you. Go back to court! Reopen the case! Make a stink!"
Enhanced   Blocked

7. Reynaldo just lost his job and is frantic about how he will pay the rent. He talks to the worker about how he will pay his rent and whether he will be eligible for unemployment. In the course of the conversation, he mentions his fear of going home to face his wife. He does not believe she will understand. The worker says, "Well, you're just upset because you're afraid of your wife and what she's going to say about this."
Enhanced   Blocked

8. Tonda is discussing her need for help with a depression that she says started several months ago. She talks about the amount of time she has missed at work and how it has reached the point that she sleeps most of the day. The worker says, "Tell me a little bit about what was going on when this all started."
Enhanced   Blocked

USEFUL RESPONSES

In this third section of the book, you will be looking at and practicing responses that enhance communication. Following are several categories of responses that you may find useful as you construct answers to the exercises in this section of the book. Learning to structure good responses is a little like learning to drive a stick-shift car, rather than an automatic one. For years your communication responses to other people have been automatic. They probably worked because you were in relationships other than professional ones. In our friendships and family relations, people communicate in shorthand. These relationships are generally positive and familiar, so others do not need to guess what it is we are saying.

Now, as a professional, you are responsible for creating an environment that makes clients comfortable and safe enough to be open. Clients will not know you very well, if at all, when they come to see you. You are in charge, therefore, of communicating in a way that makes rapport possible and collaboration on problems easy. This means that at first you are going to have to think carefully about how you are responding.

The text that follows contains openers that you can lift right off the page to get you started on the exercises you will find in the following chapters on communication. Use these initial phrases to structure constructive responses. Gradually, as you practice, you will begin to sound more like yourself. Your responses will not seem as rehearsed. At first, however, you need to practice effective ways to answer what another person has said. Using the responses provided here will help you to get started in this process.

Ways to Start Responding to Feelings

When people are talking to you about something that involves an emotion or feeling, it helps them to feel comfortable and understood if you can identify their feeling and say that back to them. When you do so, it is best to structure a single sentence and say nothing more. Anything that you might add to this could take the conversation away from where they are and over to something you have introduced. Be very careful about this. Here are some useful openers to use when you respond to feelings.

> "That must have made you feel . . ."
> "You must feel . . ."
> "You must have felt . . ."
> "That must have been . . ."
> "It sounds like you're really feeling . . ."
> "How (sad, upsetting, wonderful) . . ."
> "Sounds like you really feel . . ."
> "You must be . . ."
> "You must feel so . . ."
> "It sounds like you felt . . ."

Ways to Start Responding to Content

There are times when you might indicate you heard by responding to the content of what another person has said. In this way, you confirm that you are hearing what the client has told you and confirm for the client how important the details are to you.

> "So, it's important to you that . . ."
> "You're really concerned about . . ."
> "So you were (or he/she was) . . ."
> "Right now you want . . ."
> "So, in other words, . . ."
> "You really need . . ."
> "You're hoping that . . ."
> "It sounds like they (or he or she) . . ."
> "So they (or he or she) just . . ."

Ways to Start a Closed Question

There are times when you need facts or specific information. Questions that require only a single answer are often referred to as closed questions. Here are some ways to start a closed question.

> "What is your . . .?"
> "Where did you . . .?"
> "Who is . . .?"
> "When were you . . .?"
> "Where do you . . .?"

Ways to Start an Open Question

When we are listening to clients give background about their concerns and problems, our questions need to be more open to solicit the information the client believes is significant. Here are some ways to start an open question.

"Can you describe . . .?"
"Can you tell me a little bit about . . .?"
"Could you talk about . . .?"
"Could you describe more about . . .?"
"Can you tell me a bit more about . . .?"
"Can you fill me in on . . .?"
"Could you clarify that a little bit more for me?"
"Can you tell me something about . . .?"
"Could you say something about . . .?"

Ways to Start an "I Message"

There will be times when you are concerned about something the client has done or said. You may be worried about something the client intends to do or something the client has not done. When we looked at the roadblocks to good communication earlier in this chapter, many of the ways in which workers brought up their concerns were confrontive and superior. A better way to bring up our own concerns is to indicate that these concerns belong to us. We do this by using the word "I" first. These responses can consist of several sentences and should sound tentative rather than judgmental or decisive.

"I feel . . ."
"I'm concerned that . . ."
"I'm wondering if . . ."
"It appears to me that . . ."
"I need to understand . . ."
"I'm not clear about . . ."
"I need to kick something around with you."
"I'm having a problem with . . ."
"I'm uncomfortable that (or with) . . ."
"I guess what worries me is . . ."
"I think what I'm most concerned about is . . ."

Useful Ways to Begin a Firmer I Message

There are times when you need to act in behalf of clients. Another person may be unintentionally interfering in some way. This may call for an invitation to help, and that invitation must be worded in a way that is more authoritative, but not offensive. Here are ways to be clear about what you want or need.

"I need you to . . ."
"It would be very helpful if . . ."
"I wonder if you could help us by . . ."
"Could you . . ."
"Would you please . . ."
"We need your help to . . ."

Ways to Show Appreciation for What Has Been Said

When people bring something to your attention that is of concern to them, it is a good idea to let them know you appreciate what they have to say. Here are some ways to start appreciative responses.

"Thank you for bringing this up."
"It was good of you to tell me about this."
"I appreciate your thoughts about this."
"Thanks for telling me."
"It's helpful to me to know this."
"What might I (we) do to clear this up?"
"Can you tell me more about what happened?"

Specific Questions Useful in Beginning to Disarm Anger

When people express anger with us or our organization, it is not constructive to argue with them. Showing a genuine interest in what they are telling you is better for maintaining a good relationship. Here are some questions to ask that indicate a genuine interest on your part. By using these responses, you indicate that you really want to understand the problem the client is experiencing with you or your organization. Do not ask the person all of these questions; one or two of them will indicate a real willingness on your part to grasp the issues.

"How did I (we) offend you?"
"What did I (we) do?"
"When did I (we) do this?"
"How often did I (we) do this?"
"What else about me (us) upsets you?"
"How could I (we) do things differently?"

Examples of Ways to Agree When Practicing Disarming

When we are not acting in a professional capacity, it is common to feel very defensive when someone criticizes us. Usually, however, there is a kernel of truth in what the other person is expressing even though it might seem exaggerated or trifling to you. Here are some responses that you can use to let people know that you can see and accept the truth in what they have told you.

"I'm sure I could do better at times."
"We probably could do things a bit differently."
"There are people who have had more experience than I have."
"It may be that we could do things differently."
"Probably we're not always aware of these problems."
"It is very possible that we overlooked this."
"I certainly can be forgetful at times."
"I can see your point."

Sample Response When You Cannot Change

It may be that the person who is angry is expecting that there will be a change in the way you do things. Sometimes you are not able to change things. Maybe you are blocked by the law or how a change would affect other clients or staff. When that happens, you need a pleasant way to let the person know you cannot make those changes. Here is an example of what you might say.

> "I understand your point. We're going to have to continue this way for now, but it was helpful to hear your concerns."

Sample Response When You Find You Can Compromise

At other times, the client has brought up a useful suggestion and changes can be made. Here is an example of how you might respond in such a situation.

> "I think there are some ways we can solve this problem."

Ways to Start Collaboration

Nothing really useful can happen for the client if there is no collaboration. Even when clients will be doing most of the work, the word "we" can soften this fact and create a team approach to the problem. In this way, you let clients know they can trust your intention to be supportive without taking over and forcing a solution. Collaboration has another useful purpose. When done well, it prevents your giving the impression that you feel superior and see the client as helpless and inadequate. Here are some ways to begin collaboration.

> "Perhaps we can . . ."
> "Maybe we can (could) . . ."
> "Let's (look at this together, look at your options, see what we can find out about this)."
> "We can (could) . . ."
> "Why don't we . . ."
> "We might . . ."
> "You and I together can (could) . . ."

Ways to Involve the Client in Collaboration

Sometimes the client does not participate. You make the suggestion, and the client simply goes along with it. If this happens often, the client is not collaborating or participating in solutions to problems he or she owns. There are ways to help clients become more involved. Here are some examples.

Start with an "I Message"

> "I'm wondering if . . ."
> "It occurs to me . . ."

Finish with a Question or Comment That Invites Collaboration

> "What do you think?"
> "I'm wondering what you think."
> "What thoughts do you have?"
> "Maybe you see it differently."
> "You probably have some ideas, too."
> "But it's important to me to know how you see it."
> "But I think what would really be helpful is to hear your ideas about it."

"How do you see it?"
"What are your suggestions?"

CONCLUSION

The purpose of this chapter was to give you a sense of how constructive responses sound and how those that block rapport and understanding sound. Knowing how these responses sound and actually using them are two different things.

In each chapter in this third section of the book, the types of responses and what they are intended to do in a therapeutic sense will be discussed. Each chapter contains a series of exercises. To do these exercises, at least initially, turn back to this chapter and refer to the examples of useful responses. Use these openers to develop constructive replies of your own.

 **Chapter 9**

Listening and Responding

INTRODUCTION

Listening to others is a healing activity and, for that reason, very important in effectively helping others. Although case managers do not do therapy as we think of it in a counseling setting, they do have many opportunities to listen to others. At intake and during the course of the relationship as problems and issues arise, the case manager can offer listening as a first and an important step in resolution of the problem.

Writing in the *American Journal of Psychiatry,* Stanley W. Jackson, MD (1992) talked about the importance of listening in an article titled "The Listening Healer in the History of Psychological Healing." He wrote:

> The effective healer in the realm of psychological healing tends to be someone who is interested in talking with and listening to the other person. And these inclinations are grounded in an interest in other people and a curiosity about them. Further such healers have a capacity for caring about and being concerned about others, particularly about those who are ill, troubled, or distressed.

Describing those who seek our help, he wrote, "He seeks to be listened to, to be taken seriously, and to be understood, as crucial aspects of this process." Finally, he talked about the process itself:

> [T]he attentive listening of a concerned and interested healer can, and often does have a compelling effect on the sufferer. The sufferer often enough responds by telling more about himself, by revealing more. . . . The relationship is deepened—more is said, more is heard, more is understood, more of a sense of being understood is experienced.

Listening to another person in a way that indicates our concern for that person is important in the healing process. Active listening is a method that allows you to demonstrate that concern and interest in other people.

DEFINING ACTIVE LISTENING

Active listening is a term that was originated by Dr. Thomas Gordon to mean therapeutic listening and responding, a way of listening that is most helpful to the client. This method for listening to others has three purposes:

1. Active listening lets clients know you accurately heard their concerns and feelings.
2. Active listening creates the opportunity to correct your misperceptions.
3. Active listening illustrates your acceptance of where the client is at that moment.

When a client talks to you, there are two aspects to which you can listen and respond:

1. The *content* of what the client has said
2. The *feelings* that underlie what the client has said

Responding to feelings is empathic and is, therefore, the most useful.

When you accurately respond to the feelings the client is experiencing, the person feels heard. Someone is really listening. When clients feel that you truly hear them and the feelings they are expressing, even when they do not explicitly describe the feelings, they begin to develop trust and rapport, making it easier for them to fully talk about their problem or issue. As you saw in the chapter on roadblocks to good communication, there are many responses that are barriers to trust and rapport. For this reason, it is important to learn how to provide empathic responses that further a constructive relationship with your clients.

RESPONDING TO FEELINGS

When you let another person know that you have heard the feelings he or she is expressing at that moment, you are being empathic. Empathy is the ability to hear accurately the underlying feelings and emotions your client is expressing. Clients may tell you how they feel in actual words, but much of our understanding of other people's emotions comes from their facial expressions, body language, tone of voice, mood, and choice of words. In listening to feelings, we are listening to all of that as well as the words the clients speak. In this way, we gain an understanding of the underlying emotions and concerns.

It is important to really listen. Instead, many of us are tempted to think while other people are talking. We think about how wrong they are. We think about what they could do instead. We think about what advice we should give them and how to solve their problem.

What we really should be doing is listening for the feeling and determining the degree of that feeling. For instance, a person may be angry or furious or just annoyed. Another person may be a little wistful, sad, or openly depressed. Can you tell the difference when you listen? Of course you can, when you are really listening for the feelings.

There is a method for responding to feelings. It involves:

1. Identification of the feeling
2. Construction of a single statement that includes that feeling

This single sentence that includes the other person's feeling is called an empathic response. It is a specific way to practice empathy for the client, to acknowledge the client's feelings and concerns. It is a single sentence because to add more often distracts the other person or takes the conversation away from the central concern. A single sentence allows clients to know they have been heard and to continue on the same track, having been reassured.

Active listening does not include advice or solutions. An active listener does not tell the person to feel another way or to look at their problem from another perspective. An active listener does not judge the message or the feelings. For example, here are correct and incorrect responses to a person whose new car was damaged in a wreck:

Correct	**Incorrect**
—You must have felt pretty terrible about the accident.	—You should call your insurance dealer first before you get all upset.

To a person who just won some money in a lottery, correct and incorrect responses might be:

Correct	**Incorrect**
—That must have been thrilling!	—You may be thrilled now, but I don't think you really know what to do next. Better get a good accountant!

Sample responses to a person complaining that he did not get the correct change from the cashier might include:

Correct	**Incorrect**
—It sounds like you had some trouble with the cashier.	—Next time you go to pay your bill, try getting the manager to take your check and stay away from the cashier.

Listen to the way the worker in the following situation stays with the client throughout the exchange.

Client	**Worker**
—(Sighing as he sits down) I wrecked my car yesterday on my way to work.	—It sounds like you feel pretty bad about it.
—(Sighing again) I do. I guess I should be happy no one was hurt, but I just got the car.	—It was brand new and perfect.
—I know. I picked it out and ordered it special. It had everything I wanted. I don't know, in a way it was my fault. She ran the stop sign, but I wasn't really paying attention.	—You're sort of blaming yourself for this.
—Oh, I did. I still do. The police said she was clearly in the wrong. But now I'm going through all this unnecessary stuff with insurance and using a loaner car and trying to get a new car.	—You must feel so disrupted.
—Yeah, I do.	

Listen now as we have the same client speaks to a worker who is not trained in active listening. Notice how the worker pursues her own agenda and how the client begins to sound defensive and ultimately stops participating.

Client	**Worker**
—(Sighing as he sits down) I wrecked my car yesterday on my way to work.	—You wrecked your car? How did you do that?
—(Sighing again) I guess I should be happy no one was hurt, but I just got the car.	—Well, what happened?

—The other driver ran the stop sign, but I wasn't really paying attention. In a way, it was my fault.

—Well, you can't drive and think about ten other things. When you are driving a car, you have to pay attention to what is going on around you. If you're daydreaming, you can't do that.

—(Sounding defensive) Well, I wasn't really daydreaming or anything. I just didn't notice her. You can't always see everything the other drivers intend to do before they do it.

—I guess you can't, but I sometimes think there are just too many drivers out there anymore.

—(Nods) But now I'm going through all this unnecessary stuff with insurance and using a loaner car and trying to get a new car.

—Well, that's all part of it. You wreck your car, and you're tied up for months with all the bureaucratic paperwork. And they never give you what you need to buy another one just like it.

—Uh-huh.

Let us look at one more example of good listening. Notice how the worker in this example identifies the strongest feeling present in what the client is saying and how the client almost always responds positively to that recognition.

Client	Worker
—(Tears welling up in her eyes) I never went through this before.	—It sounds like you are devastated.
—(Nodding and crying more openly) My dad died, last week. He . . . he . . . well, he had been in dialysis, but he seemed so good Sunday night. Then Monday the nurse called me at work—oh, I guess around 10:00—and suggested I come in, but she didn't say it was an emergency or anything.	—You must have felt you didn't have to rush, that you had a little bit of time.
—Well, no. I went in as soon as I finished up what I was doing, and here he was already in intensive care. His breathing was so labored, so hard for him . . . (cries)	—That must have been a shock!
—It was! I couldn't believe it. That was my dad lying there. He just fixed my electrical outlet two weeks ago. We went out to eat for his birthday. I just couldn't believe he would go now.	—It was all so sudden.
—Oh yes, and then they are talking to my mom and I about how to make him comfortable. He knew me and all, but he couldn't talk. And you know how they say people need permission to die? Well, I told him it was okay to go (cries). I told him he gave his life a good shot and he could go. (Whispering) And he did (cries).	—That must have been so difficult for you.

In this next example, the worker's listening skills are inadequate. Notice the way the worker brings the conversation around to a more cheerful topic. In this case, the worker may very well be protecting himself from feeling the enormous pain of the client.

Client	Worker
—(Tears welling up in her eyes) I never went through this before.	—Like what?
—(Crying more openly) My dad died last week. He . . . he . . . well, he had been on dialysis, but he seemed so good Sunday night. Then Monday the nurse called me at work—oh, I guess around 10:00—and suggested I come in, but she didn't say it was an emergency or anything.	—I hope you went right in!
—Well no, I went in as soon as I finished up what I was doing. I didn't think she meant I had to hurry. And here he was already in intensive care. His breathing was so labored, so hard for him . . . (cries)	—Did he have a living will? That would have helped you to know how to handle this.
—I don't think he did. I don't know. Mom and I made the decisions. We knew he couldn't go on much longer, and we didn't want him to suffer. I couldn't believe it. That was my dad lying there. He just fixed my electrical outlet two weeks ago. We went out to eat for his birthday. I just couldn't believe he would go now.	—We never really know when death will strike, do we? We just have to look on the bright side, at all the good times we had with people while they were here. We know our parents won't live forever.
—It isn't that I thought he would live forever. It was just so sudden or something. We didn't have much warning really. I guess in that respect I can say he didn't suffer, you know, like in a long illness for years and years.	—See. There's something to be thankful for. There is a silver lining in everything.
—Sure.	

Did you notice that the client does stop crying, taking the cue from the worker that crying and continuing with the painful story about her father's death is discouraged. It must have been clear to you as well that a significant part of the story was left out. The first worker had more information about what really happened and said less. The second worker had more to say, but cut off part of the story and, therefore, did not have as much information.

Good listeners stay with the client until the emotion is drained off. If the client cries, we know we are helping that person to face and come to terms with intense emotion. A person who has been supported through this process by a skilled listener will heal better than a client who has been forced to shut down the feelings and deal with them alone.

RESPONDING TO CONTENT

When you respond to the content, you are usually doing it to check the accuracy of the information you believe you heard. Listening to content gives you clarity and helps you understand the facts.

Generally, we listen to content less often than we listen to feelings. The following are some examples of listening to content. To a person upset over having to wait for a ride, responses might be:

Correct

—So you went to the park about 7:00, and they didn't come until 4:30.

or

—In other words, the people who were to pick you up have done this to you before.

Incorrect

—Those people sound like they need a watch!

or

—You better find someone else to ride with. I can line you up with a party.

Responding to content is a good way to help a person who has just been through a traumatic event. By repeating facts of events back to clients, the clients will begin to integrate their particular experiences into the whole of their experiences, and their healing will be facilitated. In fact, the sooner a person begins this process after a traumatic event, the more readily he or she may be able to heal in the future. Here is an example of listening to content following a traumatic event:

Client

—(Looks pale and shaken, and is silent)

—(Nods) I . . . Can I tell you? It was . . . in the parking garage at the hotel. Not late or anything. I heard this person get off the elevator as I . . . (is silent).

—Yes, and as I was walking toward the car, and it was at some distance, I thought I heard the elevator and then someone started walking along behind me.

—Right. And just as I was ready to get in my car, . . . I mean, I almost made it, and he grabbed my coat and just pulled, just pulled as hard . . . just pulled me backward.

Worker

—(Responding first to feeling) It seems like you've been through something pretty terrifying.

—You had gone to the parking garage to get your car.

—They were walking behind you, but at that point you thought they were just going to their car.

—So, in other words, you were grabbed and pulled down from behind.

Notice that the worker rephrases the facts the client gives so that the client hears them again. It is not uncommon to find that through this listening process, unpleasant memories that the client has blocked begin to surface in the supportive atmosphere created by the worker. Some people would say the client is lying or making up things because her story has changed somewhat. More often, the client is beginning to remember more of what actually happened because the worker is encouraging and has created a safe environment.

POSITIVE REASONS FOR ACTIVE LISTENING

There are some specific therapeutic reasons for employing active listening.

Self-acceptance

Think of Fritz Perls and the paradoxical theory of change. The theory asserts that people change only when they are able to accept themselves exactly where they are right now.

Judgments about where clients should be only engages them in defending how they came to the place they are now. It wastes their energy and the valuable time you have to work on healing. The most healing, and therefore therapeutic, practice is active listening. By saying, "You must be drained," you accept the fact that the client is drained. If you say, "You should try to stay away from those draining people," the client now has to explain why he has not done so or cannot do so, or the client might say nothing and decide you do not really understand.

Drain Off Feeling

You will find people in all sorts of life crises and difficulties. Some of the circumstances are very traumatic, and considerable active listening will be required on your part if the client is to begin the healing process. A woman who has been raped will start coming to terms with it sooner and heal more readily if she encounters a good active listener soon afterward. An older person can prepare for the end of his life and feel comfortable about his past life if he can talk about it with a good active listener.

Human service workers have been criticized for not listening long enough. Some make only a few stabs at it, and then move into problem solving: Where will the person stay tonight? Who should be called? What facts are needed to open this case? The importance of your role in trauma, in healing, and in helping clients to grow cannot be underestimated. You do this in large part through good active listening.

POINTS TO REMEMBER

Active Listen Long Enough

Do not cut short this important piece of the client's healing process because you feel pressed for time. Be sure you go over the situation thoroughly once and review it, if you can, several times. In cases of violence, reviewing the content several times, along with listening to feelings, helps the victim begin to hear the story and come to terms with it. Active listening, in this case, promotes healing.

Solutions Come Later

Do not rush to the solution phase of the interview. Even if you have ideas, wait until the emotion has been drained off. If the client would obviously feel comforted to know there are solutions or resources, tell him or her about them, but go back and do active listening at another point in the interview.

You cannot confront the issues you feel are important if you have not done your active listening first. The client may go along with you, but not as well or with as much involvement if the person does not feel heard or understood. For instance, if you start immediately with your concerns about where the client will stay tonight instead of acknowledging the loss of her home in the fire just hours ago, she will not be as ready to work with you on solutions. Her mind is on her many losses, and her emotions may be ranging from guilt to anger. Listen first. If you help a young couple with the details of their baby's funeral without listening to their story about the baby and the baby's death, you are taking care of those details that matter most to you, but the clients are likely to experience you as unfeeling and impersonal. Your starting where they are will help them to move toward the matters that must be addressed. Most important, you are helping them to integrate this experience into the whole of their life experiences, making it easier for them to ultimately accept the reality.

Active Listening Does Not Mean You Agree

Just because you say, "You must feel very angry," this does not mean you think he should feel angry or he should not feel angry. You are simply acknowledging where the client is right now.

You Could be Wrong

You might say to the client, "It must have made you sad to see your parents go through that." The client may respond, "Well, not really. I think I felt more anger than sadness." This is a good exchange. Here you get important, corrected information that allows you to follow the client's concerns more accurately.

Mind Your Body Language

In order to facilitate the interview, lean toward the client, look the client in the eye, nod, and look interested and enthused. Do not fiddle with things on your desk, write while the client is talking, stare out the window, or glance at your watch. Give the signals that indicate you are attentive to what is being said.

HOW MANY FEELINGS CAN YOU NAME?

Instructions: In a group of no more than four people, see how many feelings you can name in twenty minutes. Remember that there are many different degrees of the same feeling.

1. _____	12. _____	23. _____
2. _____	13. _____	24. _____
3. _____	14. _____	25. _____
4. _____	15. _____	26. _____
5. _____	16. _____	27. _____
6. _____	17. _____	28. _____
7. _____	18. _____	29. _____
8. _____	19. _____	30. _____
9. _____	20. _____	31. _____
10. _____	21. _____	32. _____
11. _____	22. _____	33. _____

34. _____   57. _____   80. _____

35. _____   58. _____   81. _____

36. _____   59. _____   82. _____

37. _____   60. _____   83. _____

38. _____   61. _____   84. _____

39. _____   62. _____   85. _____

40. _____   63. _____   86. _____

41. _____   64. _____   87. _____

42. _____   65. _____   88. _____

43. _____   66. _____   89. _____

44. _____   67. _____   90. _____

45. _____   68. _____   91. _____

46. _____   69. _____   92. _____

47. _____   70. _____   93. _____

48. _____   71. _____   94. _____

49. _____   72. _____   95. _____

50. _____   73. _____   96. _____

51. _____   74. _____   97. _____

52. _____   75. _____   98. _____

53. _____   76. _____   99. _____

54. _____   77. _____   100. _____

55. _____   78. _____

56. _____   79. _____

FINDING THE RIGHT FEELING

Instructions: The important thing when responding to feelings is to know what the degree of that feeling is. It is very important to feed back to the person an accurate reading of what he or she must be feeling. All feelings have varying degrees of intensity. For each word listed here, see if you can find words that mean the same thing but indicate varying degrees of that feeling.

Example—*Happy: overjoyed, exhilarated, glad, delighted, cheerful, ecstatic, merry, radiant, content, elated, euphoric, ebullient, chipper, bouncy, bright, joyful, pleased,* _____

SAD: _____

_____

_____

_____

CONFUSED: _____

_____

_____

_____

TENSE: _____

_____

_____

_____

LONELY: _____

_____

_____

_____

STUPID: _____

_____

_____

_____

ANGRY: _____

_____

_____

_____

ACTIVE LISTENING EXERCISES

Active Listening I

Instructions: People communicate words and ideas, and sometimes it seems appropriate to respond to the content of what they have just said. Behind the words, however, lie the feelings. Often, it is most helpful to respond to the feelings. Here are several statements made by people with a problem. First, identify the feeling; write down the words you think best describe how the person might be feeling. Next, write a brief empathic response—a short sentence that includes the feeling. Refer to the sample openers provided in Chapter 8 under the heading "Useful Responses."

1. When I was in court, the defense attorney really pounded me. You know, like he thought I was lying or didn't believe me or thought I was exaggerating.

   FEELING:

   EMPATHIC RESPONSE:

2. Those dirty, lousy creeps! Everything was fine in my life, and they really, really ruined everything! I don't care if I go on or not.

   FEELING:

   EMPATHIC RESPONSE:

3. I know you said this is temporary housing and all, but I never had a place like this place. I can't stand to think I have to move again sometime, and God knows where.

   FEELING:

   EMPATHIC RESPONSE:

4. This whole setup is the pits. He gets to stay in the house after beating me half to death, and I have to go to this cramped little room. Does that make sense?

   FEELING:

   EMPATHIC RESPONSE:

5. I just can't go to court. I just can't face those guys. I will freeze up, and it will all come back—what they did to me. I just can't go through with this!

FEELING:

EMPATHIC RESPONSE:

Active Listening II

Instructions: People communicate words and ideas, and sometimes it seems appropriate to respond to the content of what they have just said. Behind the words, however, lie the feelings. Often, it is most helpful to respond to the feelings. Here are several statements made by people with a problem. First, identify the feeling; write down the words you think best describe how the person might be feeling. Next, write a brief empathic response—a short sentence that includes the feeling. Refer to the sample openers provided in Chapter 8 under the heading "Useful Responses."

1. When I was a little kid, my mom and dad got along okay, but now they fight all the time and my mother says my dad is on drugs and has a girlfriend. Home is like hell.

   FEELING:

   EMPATHIC RESPONSE:

2. I just can't go out in the car. All I hear is the screech of tires and the awful thud and scrape of metal. I thought I was dying. I can see it all before me as if it was yesterday.

   FEELING:

   EMPATHIC RESPONSE:

3. We have a neighborhood problem here! Yes we do! A real big idiot lives in that house. A real nut! Trimmed my own yard with a string trimmer and threw stones all over my car. Ruined the paint!

   FEELING:

   EMPATHIC RESPONSE:

4. I never meant to get pregnant. I know everyone says that, but I didn't! I can't think straight. What about my job and school and all my plans? I feel sick. I feel all the time like I'm going to faint.

   FEELING:

   EMPATHIC RESPONSE:

5. You don't expect us to take Alfred into our home, do you? He is very mentally ill—tore up the house several times. I really—well, I know he's my son, but I just can't deal with the way he's been in the past.

FEELING:

EMPATHIC RESPONSE:

Active Listening III

Instructions: People communicate words and ideas, and sometimes it seems appropriate to respond to the content of what they have just said. Behind the words, however, lie the feelings. Often, it is most helpful to respond to the feelings. Here are several statements made by people with a problem. First, identify the feeling; write down the words you think best describe how the person might be feeling. Next, write a brief empathic response—a short sentence that includes the feeling. Refer to the sample openers provided in Chapter 8 under the heading "Useful Responses."

1. I can tell you now, I just can't go back there. I just feel as if my husband will kill me one of these times.

   FEELING:

   EMPATHIC RESPONSE:

2. I can't stand those people! They made fun of that retarded kid night and day. I hope they get theirs!

   FEELING:

   EMPATHIC RESPONSE:

3. I've been clean for eight months! If you had told me this would happen a year ago, I'd have laughed in your face.

   FEELING:

   EMPATHIC RESPONSE:

4. Sometimes it kind of makes me sick to think of all the stuff I did when I was drinking. I'd like to go and take it all back, but how do you ever do that?

   FEELING:

   EMPATHIC RESPONSE:

5. I can tell you what scares me most. It's being by myself at the house one night and having him come back. I don't know if I can go on living there.

FEELING:

EMPATHIC RESPONSE:

Active Listening IV

Instructions: People communicate words and ideas, and sometimes it seems appropriate to respond to the content of what they have just said. Behind the words, however, lie the feelings. Often, it is most helpful to respond to the feelings. Here are several statements made by people with a problem. First, identify the feeling; write down the words you think best describe how the person might be feeling. Next, write a brief empathic response—a short sentence that includes the feeling. Refer to the sample openers provided in Chapter 8 under the heading "Useful Responses."

1. When I took that test, it was really hard. And I guess I was nervous. I mean, I couldn't think of any of the answers.

   FEELING:

   EMPATHIC RESPONSE:

2. Those lousy creeps! They're always snickering and making fun of other people, especially people who have a handicap. They make me sick!

   FEELING:

   EMPATHIC RESPONSE:

3. I know Jim said we could be buddies at swim practice, but I'm probably not as good a swimmer as he is. I feel sort of silly trying to swim with him. Maybe he would like to have a better buddy.

   FEELING:

   EMPATHIC RESPONSE:

4. This whole setup is the pits. This other guy gets the tutor, and the teacher tells me to go home and see if my mother can tutor me. She never had this math. Math isn't even her thing. Does that make sense?

   FEELING:

   EMPATHIC RESPONSE:

5. I just can't go to class. Not after making a fool of myself the last time. I got every answer wrong when the teacher called on me, and people were making fun. . . . It was terrible!

FEELING:

EMPATHIC RESPONSE:

Active Listening V

Instructions: People communicate words and ideas, and sometimes it seems appropriate to respond to the content of what they have just said. Behind the words, however, lie the feelings. Often, it is most helpful to respond to the feelings. Here are several statements made by people with a problem. First, identify the feeling; write down the words you think best describe how the person might be feeling. Next, write a brief empathic response—a short sentence that includes the feeling. Refer to the sample openers provided in Chapter 8 under the heading "Useful Responses."

1. Well, every time I go off my meds, I get kind of crazy. My minister is really putting the pressure on me to quit and let God take over my illness.

   FEELING:

   EMPATHIC RESPONSE:

2. The people at the halfway house are so nice to me, compared to the way things were with my family.

   FEELING:

   EMPATHIC RESPONSE:

3. You have some nerve, having the therapist see my son every week for six months, and then you refuse to tell me more than "he's doing better." How do I know he's doing better?

   FEELING:

   EMPATHIC RESPONSE:

4. I've been on the streets since 1972, and I never slept inside a night until now. I don't know, I just can't seem to stay out like I used to without getting this cough.

   FEELING:

   EMPATHIC RESPONSE:

5. I worry about my daughter living with that man. At first we tried to accept him, but this constant abuse and denigration . . . she wasn't raised like that.

FEELING:

EMPATHIC RESPONSE:

 **Chapter 10**

Asking Questions

INTRODUCTION

Sometimes when we are listening to another person's difficulties, we might ask a lot of questions. We are usually well meaning. In part, we do this to find out more so a solution can be quickly devised. We also do this as a way of filling in the gaps when our active listening skills are not strong. It may be easier for the listener to say "Did you have any money?" or "Where did the man say he lived?" than to simply say, "That must have been so hard on you."

Sometimes when we listen, we feel nervous about what the other person expects from us. After all, if we are the worker, should we not have all the answers? If we do not have an answer just yet, we can stall for time by asking a lot of questions until we do.

Often, however, questions are received by the other person as prying. You may ask questions at a rate the clients are not ready to answer. For instance, you may be asking questions further ahead in the story, throwing the client off. Clients may feel pushed to reveal more than they intended or distracted from the line of reasoning they were following. They may become defensive and their communication guarded if the questions seem to pry or to imply there is only one way to have handled things.

WHEN QUESTIONS ARE IMPORTANT

Obviously, at times questions need to be asked. The three times when it is important to ask questions are:

1. When you are opening a case or chart on a person and need identifying information
2. When you are compiling information for assessment and referral purposes and need facts to do that properly
3. When you are encouraging the person to talk about his or her situation freely in order to better understand which aspects of it are important to him or her

CLOSED QUESTIONS

A closed question is *one that requires a single answer.* For example:

Worker	Client
— Where do you live?	— 346 Pine Street
— How long have you lived there?	— Oh, about 6 years
— Have you ever been seen here before?	— Yes, in 1998
— And who did you see then?	— Dr. Langley in outpatient
— Did she prescribe any medication?	— Yes, she gave me a prescription for Prozac.
— Do you still take that?	— No, I stopped a few months ago.

The times you most often use closed questions are:

1. When you are opening a case or chart on a person and need identifying information
2. When you are compiling information for assessment and referral purposes and need facts to do that properly

In both of these instances, however, clients will need to talk about what has brought them into the service, and you will not rely exclusively on closed questions. However, closed questions are appropriate here to gather the basic information. The client can respond with a simple answer. Expressions of feelings, descriptions of circumstances, and explanations for problems are not being asked for.

OPEN QUESTIONS

Open questions serve the purpose of giving clients more opportunity to talk about what is important to them. By asking open questions, you receive more information about a client's situation. In answering open questions, the client can talk about feelings, underlying causes, supporting circumstances, and personal plans.

Open questions have been shown to put clients at ease. Workers using these questions are not perceived as prying, but as expressing real interest in other people or a genuine desire to understand their situation. You can use open questions to obtain examples or elaboration of the problem and to clarify certain aspects of the other person's story.

Often an open question begins with "can" or "could," but there are other ways to start such questions. For example, you might say, "Tell me a little bit more about your divorce." Typically open questions look like those in the following example.

Worker	Client
— Can you tell me about the night your father left?	— Well, my mother had been arguing with him for some time. I could tell he was getting angry. I don't think I really blame her for his leaving. He had done many things to her that she had every right to be angry about. But I guess for him it was the last straw. Anyway, we were having dinner and she began on the same topic of the house. He just put down his fork and got up from the table and walked out of the house.

— Can you describe your relationship with him after that?

— Well, he did come back to the house for his things from time to time. He got an apartment nearby, and I used to stop there on my way home from school. We never stopped seeing each other, and we never talked about my mother. I can't ever remember him asking me how she was or, for that matter, saying anything mean about her.

Would the worker have gained as much information by conducting the same conversation using closed questions? This example demonstrates that the client would have been much less forthcoming.

Worker	Client
— When did your father leave?	— Oh, August of 1995
— Did your parents argue much before he left?	— Sure, yeah, a lot of the time
— Why did he go?	— Well, they disagreed over money.
— After he left, where did he go?	— He got an apartment near us.
— Did you ever see him after that?	— Yes, pretty much

Would the client have felt investigated with these short closed questions fired at him, one after the other? Did the worker really understand how the client felt about the divorce and the contact with his father?

In the second example where the worker used closed questions, there was room for the worker to assume things that might not be true. For instance, the worker might have assumed that the client blamed his father for leaving. In the first example where the worker used open questions, this clearly was not the case.

QUESTIONS THAT MAKE THE CLIENT FEEL UNCOMFORTABLE

Avoid the Use of "Why" Questions

If you ask someone why they did something or did not do something, you imply that you believe the person should have handled things differently.

Why didn't you call the police?
Why did you go there?
Why were your children out that late?

Do Not Ask Multiple Questions

If you fire off a string of questions, the person can feel interrogated. You may sound impatient, and you can confuse the client.

Did you see the other person? Did you turn right around then and call the police about it? Did you get a license plate number? Were you standing close to the window?

Do Not Change the Subject

If the client is talking about how she learned of her mother's death, do not start asking questions about her mother's prearranged funeral. Let the client continue to talk about her mother's death until it seems that she wants to turn to the prearranged funeral. Never ask about something out of curiosity. Do not ask, for example, about the prearranged funeral because you are thinking of getting one for your mother and want to know more about it. Ask questions that stay on the topic the client has selected; if you do ask questions on another topic, be sure it is relevant and will actually clarify the client's situation for you. Here are some examples of questions that change the subject.

> I heard about the kids before, but where do you work?
> So, she died on Saturday, and now you are seeing a lawyer about the will?

Sometimes questions such as these can disrupt the entire discussion.

> That's real neat about how your car looked before the accident. Do you have adequate car insurance?
> So, your mother died on Saturday, and you're living in a house by yourself?

Do Not Imply There Is Only One Answer to Your Question

You can ask questions in a way that implies there is only one thing the client should have done or thought or said.

> Didn't you go to the police?
> Did you *tell* the other person what you heard?
> Did you see to it that he knew what you were thinking?

Do Not Inflict Your Values on the Client

You can also ask questions based on your own value system. The client, however, may have other values. For example, you may value truthfulness at all costs, while your client may come from a group that values group harmony and not hurting another person's feelings. Questions that imply that your value system is better are not useful.

> Did you tell her how you felt before you just walked out?
> Did you tell her the complete truth?
> Don't you value truth above everything else in this situation?

Do Not Ask Questions That Make Assumptions

You can word the question in such a way as to make it clear that you are assuming you already know the answer.

> You called the police, right?
> You wanted to go to the store, didn't you?
> He was being a fool, wasn't he?

A FORMULA FOR ASKING OPEN QUESTIONS

Figure 10.1 contains a formula for asking open questions. In the figure, the open question is broken into parts. You can interchange the parts, by choosing one part from each column, to construct good open questions that encourage the other person to feel safe in talking and expressing feelings and opinions. Use this formula in the exercises that follow to construct effective open questions that invite others to be open and talk freely with you.

WHAT IS WRONG WITH THESE QUESTIONS?

Instructions: Look at the questions that follow and decide what makes them bad questions. In writing your criticism, look for questions that assume there is only one answer, inflict values on the client, make the client defensive, make assumptions, cut off discussion, or change the subject.

1. A woman is telling a worker about why she has come to the shelter tonight. Right in the middle of her gripping tale about what was going on at home only a few hours before, the worker says, "How long has this been going on?"

2. A worker has listened to a young mother talk about how she dropped out of school and got pregnant and has no skills. Finally the worker interrupts to ask, "Did you have to get pregnant? Didn't you know about birth control?"

Formula for Asking Open Questions

Openers	Directives	Add-ons/Softeners	Object of the Question
Can you	share	a little bit more about	your husband
	describe	a little bit about	your childhood
	explain	a little more about	your medication
	summarize	something about	what the move was like
	outline	the problems with	the move
	spell out	the larger picture	regarding the move
Could you	talk	a little more about	what your dad said
	give me	a bit more about	your illness
	tell me	something more about	your job
	help me understand	something about	your relationship with your kids
	clarify	a bit	the situation

Figure 10.1 Formula for open questions

3. A man calls and says he is depressed. He has felt depressed for some time and is now thinking of suicide. The worker asks, "Where is your wife? Are you divorced?"

4. A man is telling you about the night he witnessed a murder. The victim was his brother-in-law, and although he was never very close to him, he feels that maybe he could have stopped his death in some way. The worker asks, "Why don't you just go and ask the police?"

5. A woman has come into temporary shelter with a lot of debts. She has been out looking for work today and is discouraged about not finding anything yet. She sits down tiredly in the worker's office and tells about what her day was like. The worker asks, "Did you have to get so many debts?"

6. A man wants to know if his wife is all right after she has been raped. He is sitting with a worker in the waiting room while his wife is being seen in the emergency room. The worker answers his question with one of her own: "How much does your wife mean to you?"

7. A patient in a partial hospitalization program for the chronically mentally ill tells the worker that when the group went to the mall, one of the patients took a pair of socks without paying for them. The worker asks, "You told someone right away, didn't you?"

8. A woman is telling about the time her coworkers waste when the supervisors are out at meetings all day. The worker responds, "Why don't you say something?"

9. A woman tells a worker about a long and difficult marriage she has endured. She mentions abuse, both verbal and physical, and talks about her own failing health in

recent months. The worker asks, "Why can't you just bring yourself to divorce him?"

10. A man is trying to sort out whether or not to leave his employer. He feels that the small company is poorly run and that he could do a better job if he went out on his own. On the other hand, he likes his employer; and he feels sorry for him and the mess he's made of his business. He knows that if he leaves, things will really fall apart. The worker asks, "Don't you value loyalty?"

WHICH QUESTION IS BETTER?

Instructions: Look at the following questions and decide which of them are better than others. Place a check mark next to those you think are good questions, and then explain why you think they are better than the ones you did not check.

1. The worker to a woman in the hospital waiting room whose baby just died of pneumonia: "How old was your baby?"

2. The worker to a woman who is grieving after her husband died in a hunting accident: "Could you tell me about your husband?"

3. The worker to a teenage boy who is afraid of failing a math course and losing an opportunity to get a scholarship: "Can you tell me a little bit about this math course?"

4. The worker to a young woman who has just discovered her best friend and her boyfriend have been seeing each other behind her back: "Can you tell me something about your best friend?"

5. The worker to an elderly woman whose dog of fifteen years has died: "Couldn't you get another one?"

☐

6. The worker to a man who is requesting food for his family after running out of unemployment compensation and being unable to find a job: "Can you describe the sort of work you would be looking for?"

☐

7. The worker to a woman in a shelter who has been out searching unsuccessfully for a house or apartment for herself and her two children: "Where all did you look?"

☐

8. The worker to a single mother who has been referred for parenting skills training: "Could you tell me something about the problems you have been having with Johnnie?"

☐

9. The worker to a man with developmental disabilities whose mother, with whom he has always lived, died unexpectantly: "What did your mother die of?"

☐

10. The worker to a woman who was accosted and assaulted in her neighborhood and is afraid of calling the police: "Can you tell me a little bit about what happened tonight?"

☐

EXERCISES IN OPENING CLOSED QUESTIONS

Opening Closed Questions I

Instructions: Following are some vignettes in which the worker asks closed questions. Write an open question you think might work better in each situation, and be prepared to tell why you think the closed question is not useful.

1. A human service worker in the emergency room is talking to a man who was hit on the head before he was robbed. He seems to be having trouble getting the story out, but he wants to tell the worker everything that happened. The worker thinks that it is late and that the man ought to get to bed. The worker cuts off the discussion with, "Aren't you tired, Mr. Jones?" What open question would you have asked Mr. Jones to help him wrap up his story?

2. The human service worker is trying to learn what happened that resulted in Mrs. Peters being without housing. Mrs. Peters says she has been "on the street awhile now." The worker asks, "Have you been on the street for two years, three years?" What open question would you have asked to learn more about what happened to Mrs. Peters to make her homeless?

3. The human service worker is on the phone with a woman, the victim of child abuse. The woman tells how she has felt recently, how she needed to call, and then sighs and says, "Oh, I don't know how to begin." The worker asks, "Did your father do this to you?" What open question would you have asked that would have helped the woman to start telling the story in her own way?

4. A child is talking to a youth worker while he waits for his mother to get a place to stay. "We've lived in sixteen places," he announces, "and I'm only 7." The worker says, "What school did you go to last?" What open question would you have asked to help the child talk about what all this moving has been like for him?

5. A man calls a hot line and tells the mental health worker he wants to die. The volunteer asks, "Does this have to do with being abused as a child?" The man is startled and says, "Why, uh, no. Not really." The worker asks, "Well, what's the problem?" What open question would you have asked to help the man to talk about what was troubling him?

Opening Closed Questions II

Instructions: Put yourself in the place of the worker in the following vignettes, and decide what question you would have asked in each situation. Write an open question that you think might work better than the one asked by the worker, and be prepared to tell why you think the closed question is not useful.

1. A worker is interviewing a man in the food bank. He tells the worker that he and his children have not eaten for twenty-four hours and that he has spent most of that time getting referred around town until he finally got a voucher to come to you for food. The worker asks, "Why don't you have any food?" What would you ask?

2. A woman is referred to the social service department in a large hospital after having a stroke. She is somewhat incapacitated and has had a lot of therapy while hospitalized. Now she is going home and needs therapy at home. The worker asks, "What kind of therapy do you want?" What would you ask?

3. A man and woman have been referred by the county Children and Youth Services for parenting skills training. They are poor and have had their four children removed from the home. They have been told the children will be returned when they complete the course and demonstrate they can use the skills they learned in supervised visitations. The worker asks, "Are your children good kids?" What would you ask?

4. An elderly woman has been having trouble caring for herself in her own home. Twice now, in the middle of the night, she has called an ambulance and has been taken to the hospital for chest pains. When her heart is checked, she is found to be in good health, if a little frail. The worker who is looking into what could be going on asks, "Are you afraid to stay at home alone?" What would you ask?

5. A young woman and her baby have been given a voucher for temporary shelter after she lost the apartment in which she was living. She was evicted for back rent, and her rent fell into arrears only when she was laid off several months ago.

She has worked, but she cannot earn quite what she was making before. The worker doing the intake interview asks, "What kind of work have you been doing?" What would you ask?

Try Asking Questions

Instructions: Look at the case histories that follow and, for each one, write four closed questions and four open questions that you might ask the client.

1. Annette came to your office needing her prescription filled. She was in Marywood Hospital, a private mental hospital, and was discharged on Tuesday. She was given prescriptions, but has no money to fill them. She has no job and probably is eligible for prescriptions paid for by the county. You open a case on her.

   YOUR CLOSED QUESTIONS TO OPEN HER CASE ARE:

   1.

   2.

   3.

   4.

   YOUR OPEN QUESTIONS TO LEARN MORE ABOUT HER ARE:

   1.

   2.

   3.

   4.

2. Marie was a client of a partial hospitalization program. She was loud and demanding, but often felt hurt upon learning that others were afraid of her or reacted to her as if she was being angry. As a result of an encounter in the partial program, she is sent to you, her new case manager, to see if there are ways to help her that might work better. You need to understand more clearly what has happened from her perspective and what sort of program she might fit into.

   YOUR CLOSED QUESTIONS TO BECOME ACQUAINTED WITH HER CASE ARE:

   1.

   2.

   3.

   4.

YOUR OPEN QUESTIONS TO LEARN MORE ABOUT HER PROBLEMS AND
DESIRES FOR TREATMENT ARE:

1.

2.

3.

4.

 **Chapter 11**

Bringing Up
Difficult Issues

INTRODUCTION

There will be times when you have a concern about something the client has said or done. Often this will be because you are concerned for your clients and do not want to see them do something harmful or continue to behave or think in ways that are destructive. Occasionally you will have a problem because the client has in some way interfered with your ability to do your job well. As we noted earlier, when your needs are not met, you are responsible for resolving the matter or, at the very least, for bringing your concerns out in the open where they can be discussed and examined by the client.

Bringing something out into the open is called confrontation. To most people this means an angry, accusing action. In social services, however, it means bringing something out to gain a better understanding and perhaps to make meaningful changes or take important new steps. Many opportunities to grow and make constructive changes will be discovered when you use confrontation.

The decision to use confrontation is another strategic decision. There are guidelines as to when confrontation might be a useful tool to help you and your clients explore differences and resolve possible conflicts.

WHEN TO USE CONFRONTATION

Discrepancies

There are times when the client will communicate two different messages. Some examples follow. Confrontation can help the client to see these discrepancies and can offer an opportunity to look at the situation and at the person in another way.

The client says one thing but does another. Dalia tells you that she really wants to go to the job training program and that getting a job is a top priority for her. On the other hand, she does not go down to register for the classes because of numerous excuses, some of which do not seem entirely believable.

The client has one perception of events or circumstances, and you have another. Harold thinks you are uncaring and self-involved. He got this idea because you did not come to work the Friday after Thanksgiving even though the office was open.

He was off work that day, and he wanted to make an appointment with you so that he would not have to miss work at another time. Your perception is different. To you it was reasonable to be off work the Friday after Thanksgiving because there was only a skeleton staff working that day. You also needed to take a day off before the end of the year or you would have lost some of your accumulated time. Clients rarely come in on this date, and there was a crisis team to cover any crisis that might have come up. To Harold you seem uncaring, while to you your actions seem reasonable.

The client tells you one thing, but her body language sends a very different message. Andrea may tell you that she is "fine," that she "feels okay," or that "everything is all right." She looks, however, as if the opposite is true. She may speak in a monotone, look at the floor as she speaks, and appear depressed or disheveled. These are clues that the spoken message and the unspoken message do not match.

The client purports to hold certain values but violates those values by his behavior. Paul tells you he "likes everyone" and how he "accepts" everyone. He tells you ethnic differences are unimportant to him, and he finds them enriching. In one of his meetings with you, he tells a decidedly racist joke that obviously denigrates an ethnic group.

Other Reasons to Use Confrontation

There are other reasons for bringing out in the open behavior or communications that seem to interfere with the client meeting his or her goals.

The client has unrealistic expectations for you. Marcy expects that you will drop everything to see her or to take her phone calls. She does not want to see anyone else in the agency and does not think she should have to see anyone else at night. You are her case manager, and she wants you to be there when she needs you.

The client has unrealistic expectations for himself or herself. Miguel has been in a partial hospitalization program for a number of months and has been sick for about four years. Stress seems to trigger his schizophrenic symptoms, and regulating his medication is difficult. He is very good at cleaning and janitorial tasks around the center, and there is a good supervised janitorial program for clients where they hold a regular job and clean actual establishments. Miguel is set on going to work at the highway department and getting a job driving a steamroller. He applies for the job repeatedly but gets no response.

The client asks you for assistance, but her actions indicate she is not interested. Serena asks you to help her find suitable housing so she will not have to stay at the shelter any longer. You have some leads she could pursue, but she breaks appointments, calling in to say she was detained and will reschedule. She does not follow up on the leads you give her, and the two apartments she went to see that were suitable she turned down for minor problems, refusing to live there.

The client's behavior is contradictory. Art comes in to group and tells the group he will stop drinking. He never misses AA meetings, gets a good job, and begins to help others to stop drinking. Later you learn that he is actually drinking in spite of what he says in group and at AA meetings and that he goes to AA on Tuesday and Thursday and to his favorite bar on Friday and Saturday nights. Art's behavior is contradictory in another way. While he talks to newcomers in the group about how helpful it is to stay in group and how wonderful the agency is, he has been denigrating a certain staff person outside the building where he goes to smoke during the break.

THE "I MESSAGE" IN CONFRONTATION

Because the problem is yours and the observations are your own, confrontations should begin with or include a reference to you. The term used for these statements by Dr. Thomas Gordon was "I messages" because they contain the words "I" and "me." Confrontation is not helpful, as we have seen, if it uses the accusatory "you." The following examples of messages to a client who was late on Tuesday demonstrate the difference:

Correct	Incorrect
— I'm concerned that you were late on Tuesday morning. It got my day behind more than I wanted, and I spent a lot of time trying to catch up.	— You were late on Tuesday, and you held me up. My whole day was behind, and I spent a lot of time trying to catch up because of you.

A complete "I message" usually contains four parts:

1. Your concerns/feelings/observations about the situation
2. A nonblaming description of what you have seen or heard—of the behavior
3. The tangible outcome for you as a result or the possible consequences for the client
4. An invitation to collaborate on a solution

In the following examples, messages given to clients are broken into the four parts. Compare the correct and incorrect messages:

Correct	Incorrect
1. I'm concerned that . . .	You were . . . (There is no "I" in this message. It starts with the accusatory "you".)
2. Your being late for the last four appointments has	You were late for the last four appointments and
3. Caused a lot of scrambling on my part to catch up the rest of the day.	You caused a lot of scrambling on my part to catch up the rest of the day.
4. Can we look at what is happening here?	(There is no invitation to collaborate on a solution in this message.)

Correct	Incorrect
1. It appears to me that you could get in trouble . . .	If you . . . (There is no "I" in this message. It moves directly into the accusation.)
2. if you follow through on your plan to yell at the District Justice.	go out there and yell at the District Justice,
3. It might cause him to be even tougher on you.	all you are going to do is get yourself in a lot of trouble.
4. Let's look at this.	My advice is to cool your jets. (There is no invitation to collaborate. The worker gives advice instead.)

THE RULES FOR CONFRONTATION

There are ways to talk with someone about the issues that concern you. An important goal is to do so in a way that allows the client to hear you and make use of what you have said. We all benefit from the feedback of others, but the manner in which it is given often interferes with our ability to accept and use that feedback.

Note that in the "correct" examples provided in this section of text, the speaker emphasizes "I" and "me," taking responsibility for the observations and concerns. In the "incorrect" examples, the emphasis is on "you."

The following text discusses the rules for making "I messages" less threatening and more acceptable to the listener.

Be Matter-of-fact

Do not become excited or judgmental or petulant. For example, to a person whose goals are unrealistic for the present, correct and incorrect messages would be:

Correct

— I need to talk with you about something that bothers me. It seems to me that some of your goals are a bit further down the road. Perhaps we could look at some preliminary steps for you to take first to help you get ready.

Incorrect

— You better reconsider! You're not at all ready to undertake a job like that. Let's get cracking on some training first, something to prepare you. You don't just walk in and get the best job right away.

Be Tentative

You could be wrong in your observations. For that reason, it is not helpful to present yourself as though you know everything. These messages to a person who may not be seeing all of the issues with his mother demonstrate the difference:

Correct

— I'm wondering about this problem you're having with your mother. I could be wrong, but when you describe the way she talks to you, it sounds as if she is angry for some reason. What do you think?

Incorrect

— Your mother is obviously angry at you!

Focus on Tangible Behavior or Communication, What You Can Observe

Sometimes when we bring something up for discussion, we tend to be vague about what the actual problem is. We might generalize or just describe our feelings about it. This is not enough information for the client to use to make a meaningful change. The correct message in the following examples does provide such information to a client who is frequently late for appointments:

Correct	**Incorrect**
— I have a problem with the number of times you have come for your appointment late. Maybe we should take a look at it together and see if you can make some arrangements to be on time. For example, you were 20 minutes late on July 10th, 1 hour late on July 17th, and 45 minutes late on July 24th. I need to talk about what is happening here.	— You're always late. Every time we have an appointment, you come in when you feel like it. or — I'm really upset with you. You're never on time.

Take Full Responsibility for Your Observations

If you recognize that what you observed is what *you* observed and that it is perfectly all right for your observations to be incorrect, it will be easier for you to take responsibility for your observations. If you are wrong, the perception can be corrected, particularly if you have been tentative and nonjudgmental. Notice the differences in these messages to a person who needs housing but is doing little to obtain it:

Correct	**Incorrect**
— It appears to me as if the sessions we have together to get you better housing aren't as important to you as I first thought. Let me tell you what I mean. To me it seems you have other more important priorities. I'm basing this on the fact that you never went to see the three apartments that were available to you. Can we talk about where you are right now with this?	— I can see you don't care about housing. or — The way it appears, housing certainly isn't a high priority for you! You never follow through.

Always Collaborate

Share responsibility for finding a solution or an understanding.

Correct	**Incorrect**

— How can we look at this differently?
  or
— What can we do to change this?
  or
— Is there something we should be doing differently?
  or
— How can we resolve this?
  or
— Let's look at this together.
	— You better do things differently.
	  or
	— You need to change things.
	  or
	— I hope you can figure out how to handle this thing.
	  or
	— You need to find a solution here.
	  or
	— You better take a good look at this yourself.

Do Not Accuse the Other Person

It may be tempting to blame or accuse the client for the situation. Refrain from doing that because it prevents the client from hearing you, as demonstrated by the following examples of messages to a person who is frequently late:

Correct	Incorrect
— When you come late, my appointments get backed up, and the entire day is thrown off schedule.	— You have a punctuality problem, don't you? My day is all thrown off because of you!

Do Not Confront Because You Are Angry

Sometimes it is tempting to use confrontation to punish a client who has made you angry. In these situations, you might use public humiliation or denigrate the person as a person. Again, the client will not hear the important message, as you can see from these examples of messages to a person who is having trouble maintaining sobriety:

Correct	Incorrect
— I'm concerned that you are drinking when you are away from the program and not talking about these relapses in the group. I think to me it seems untrustworthy not to be honest in group.	— You can't come to group and lie to people about your drinking. You've been drinking outside group and you're lying about it. Plain and simple, it looks to me like you just plain lied to us.

Do Not Be Judgmental

Do not sit in judgment of the client, as the worker does in the incorrect example that follows. These are messages to the person who is not pursuing permanent housing:

Correct	Incorrect
— Can we take another look at your priorities and see where housing for you and your children fits in. I was under the impression that this was pretty high on your list, but you haven't kept the four appointments we had to discuss it.	— If I were you, I'd make housing a top priority. You have two children, you're living in a shelter, and you aren't doing a thing to change the situation. That's what I call irresponsible.

Do Not Give the Client a Solution

Because of your position with regard to the client, who is already having problems, any solution you give will be seen as imperative. We want clients to develop their own solutions. Even the words "should" and "ought" sound like imperatives to the client and are best avoided. These are examples of messages given to a person having trouble remembering appointments:

Correct	Incorrect

— Let's see if there is a way to resolve this.
    or
	— Go get an appointment book. Write all of our appointments in the book, and that way you won't forget.
	    or

— There probably are some different ways we could approach this. I have some thoughts, and you probably do too.	— You should get an alarm clock that works and have your landlady call you up every morning. That way you can't miss.

CONFRONTING COLLATERALS

There are times when someone is interfering with the client's treatment or your ability to interact effectively with the client. For instance, some years ago a night-shift nurse supervisor in the emergency room took it upon herself to keep the interview room open. Even though the room was there for workers to interview victims of domestic violence, violent crime, or rape, the nurse would barge in, in the middle of the interview, and try to clear the room.

Such situations generally include something someone is doing that:

1. Adversely affects the client
2. Adversely affects your work with the client

In situations like this, you need a firmer message. The message would:

1. Not sound tentative
2. Be pleasant, but firm (Smile, but mean what you say.)
3. Contain an implied or explicit request for help

Correct and incorrect messages to the nurse in the emergency room who is trying to clear the interview room might be:

Correct	Incorrect

— I need you to help me complete this interview. I expect to need this room for about 45 minutes, and then I will have all the necessary information.
 	
    or
 	
— Could you give us another 45 minutes to complete the interview? This must be done before the patient leaves.
	— Oh dear, we'll only be a minute, and I need this information too. May we stay awhile longer?
	
	    or
	
	— I thought we could use this room any time. What seems to be your problem?

In confronting other people, it is tempting to throw out the rules and simply show our annoyance or exasperation. The problem with that approach lies in the fact that you need to work with other people and the agencies they represent. In this field, we must be able to communicate well with one another if we expect to help the people we serve learn better ways of communicating. Your anger directed toward the nurse in the emergency room can affect relations between your agency and the entire emergency room staff. If this is an important part of your work, such strained relations will affect patient care. Remaining firm, but diplomatic, often prevents that problem.

ON NOT BECOMING OVERBEARING

It is a little tricky to stay where the client is and still express your own concern. Sometimes a technically correct "I message" is really about your agenda and is not sensitive to the client and where the client is with the problem at the moment. Such an "I message" comes across as intrusive.

For instance, a woman is suddenly widowed. Her husband died in an accident on Tuesday night. You went to the home as part of the crisis team the night it happened because police said she was extremely upset. Tonight you are doing a follow-up visit. When you talked to her the first time, you learned that she is the second oldest of five children. Her brothers and sisters do not live nearby, and she made no move to call them in spite of your suggestion that she do so and your offer to do it for her. You feel that family can be very supportive at a time like this. You have reached this conclusion because you and your family are close and supportive. In this situation, you might send "I messages" like those that follow. The parts that are italicized actually express a view or opinion belonging to the worker and do not leave any room for the client's perception.

> "I will honor your request; however, *I feel you may be avoiding a source of real help.*"
> "I'm uncomfortable that you don't want your family to be aware of your husband's death. *Family support can be very comforting, and I'm sure that they will not be inconvenienced.*"
> "*I'm not clear* about why you want to keep this from your family. *I feel that they would want to know.*"
> "*It seems to me* that going through this alone *will be very rough for you.*"
> "*I feel that talking to your relatives will be very helpful.*"
> "*I have a problem with you wanting to do this alone.*"

Suppose it turns out that some years ago, this woman was in trouble. She was a rebellious teenager and left school and ran away from home. Her parents seemed not to care, and when she attempted to return home at age 19, they told her she had caused them enough grief and she was not welcome there. She moved here, went to college, got a master's degree, and married a local dentist. She feels better off without her family who has never offered her support in the past. She does not tell you all this because she just met you and she does not know you well enough to go into all the reasons why she left home and is estranged from her family.

Think about such possibilities very carefully when you frame an "I message." Be sure that while you speak your concern, you leave plenty of room for the fact that you do not know everything and that you could be very wrong. Sounding tentative helps.

WHAT IS WRONG HERE?

Instructions: Look at the following confrontations, and identify what is wrong with the way each one is expressed.

1. To a person who is drinking and taking tranquilizers: "That's a dumb thing to do!"

2. To a person who is driving without a driver's license: "You're just doing this to tempt fate."

3. To a person who is always forgetting to take his insulin: "I'm sick of these so-called lapses of memory. You must want to feel sick most of the time!"

4. To a person who bounced three checks in three months because she cannot seem to balance her checkbook: "Go take an accounting course, for heaven's sake!"

5. To the person who has lamented not spending enough time with his son: "Children are important. They grow up fast. You only have so long to spend with them when they are kids. You need to see that."

6. To the person who had trouble completing a high school equivalency exam and is now talking of becoming a doctor: "You need to be more realistic about what you can and can't do. You need to think of something else."

7. To the woman who has completed ten weeks in a rape victim support group and is still unable to work or leave the house much, but who says she is fine and getting over it: "It doesn't seem to me like you're getting over it. If you wanted to get better, you would force yourself to go out more."

8. To the man who complains about his neighbors but spends time on his porch yelling at the children, which starts neighborhood feuds and tensions: "You're always yelling at them. Of course they fight with you!"

9. To the woman who has been in a wheelchair for several months following an accident in spite of her doctor's feelings that she could now be up walking with

crutches: "You need to get out of that chair and practice walking. Obviously you get something out of sitting in that wheelchair."

10. To a child who says the other kids do not like him, but who is always hitting the other children and calling them provocative names: "You're half the problem, you know. Stop yelling and hitting everyone, and they'll like you better."

EXPRESSING YOUR CONCERN EXERCISES

Expressing Your Concern I

Instructions: In each of the vignettes that follow, you have a problem—a concern about something affecting the consumer. For each of these situations, construct an "I message" from you to the consumer. Be sure to follow the rules for confrontation.

1. A woman, who has been the victim of grave physical abuse, is currently staying in a shelter where you see her. One night she comes in drunk and tells you, "Hey, it doesn't hurt as much this way." The next day you approach her with an "I message" expressing your concern.

2. A woman calls and says her husband is really a dear. He has done many wonderful things for her, and she is feeling guilty about calling you, but he does keep her confined to the house and slaps her a lot. You use an "I message" to express your concern.

3. A man with two children needs temporary shelter. His oldest, a daughter, is old enough to drop out of school; and in the course of placing him, you learn that he has encouraged her to do just that. He tells you he needs someone at home to look after the place, now that they have one, and to see that the younger child is taken care of. You use an "I message" to express your concern.

4. A woman, the victim of a violent crime, is using a prescription medication her doctor gave her to help her with the anxiety of facing the perpetrator in court. Lately you feel she has been abusing her medication. Her speech seems slurred, and you often see her slip one of the pills into her mouth. You use an "I message" to express your concern.

5. A woman has not come out of her house since she suffered a major injury at work. Although her doctors say she will be able to return to work if she goes to rehabilitation, she refuses to go and cites her concern for her fragile recovery. You have talked to her many times by phone and invited her to attend support groups at the rehabilitation center where you work and to see a counselor, but she never comes, and you are becoming aware that she is terribly fearful. You use an "I message" to express your concern.

Now return to the first exercise called "What Is Wrong Here?" and construct better "I messages" for each situation described there.

Expressing Your Concern II

Instructions: In each of the following vignettes, you have a problem—a concern about something affecting the consumer. For each situation, construct an "I message" from you to the consumer. Be sure to follow the rules for confrontation.

1. A man who has been sitting by his wife's side since she slipped into a coma is weary and has neither eaten nor slept for over twenty-four hours. You approach him with an "I message" expressing your concern.

2. A woman who is refusing to take medication that would prevent her from having a psychotic episode comes to you and says she is not sure what to do. She does not feel well, but she would like to be able to handle things without medication. You use an "I message" to express your concern.

3. A man whose wife just left has told you he wants to give up his job and simply leave the area, having no further contact with either his ex-wife or his children.

You are concerned that he has not had time to think this through. You use an "I message" to express your concern.

4. A woman staying in the shelter where you work has left her baby in the care of others repeatedly and gone out. She says she is going to the store or to look for an apartment or a job, but others let the baby lie in the crib and cry. You have had to feed and change the baby on several occasions. You use an "I message" to express your concern.

5. A man is waiting for his Social Security disability check to start. He has a serious heart condition and has been told he should not be out in extremely cold or hot weather. You stop by on a home visit and discover he is out on a cold day shoving piles of snow off the driveway. He tells you it is not that cold and this is not "shoveling." You use an "I message" to express your concern.

EXPRESSING A STRONGER MESSAGE

Instructions: In each vignette that follows, you have a problem with the behavior or actions of someone; this person's behavior is affecting the goals of your work with the client or is adversely affecting the client. For each vignette, construct a firmer message that explicitly or implicitly requests this person's help.

1. You are interviewing a man who appears to be quite delusional in the hospital emergency room. The new security officer at the hospital does not seem to understand that the behavior is part of an illness, and he keeps entering the room and asking, as if the patient cannot hear, "Is he giving you any trouble? Do you want me to take care of him?" Your message expresses your need to continue the interview and your need for privacy.

2. You have been working with a man who was beaten and robbed. Because of the injuries, he has been unable to work. His employer calls you several times, saying he thinks the man is simply "freaked out" and needs to get over it. The boss tells you that he has told the man this on several occasions, and says that the man just yells at him. You need the boss to understand the severity of the situation, and

you feel it would be helpful if he did not keep calling the victim with his negative opinions. Your message expresses your need for the boss to work with you and the client more constructively.

3. You have been working with a child in temporary housing. You have discovered the child is very artistic, and you have found an artist who is willing to volunteer time to teach the child on Saturday mornings. The mother of the child is upset and tells you that it is impossible "the kid has any talent" and that "anyway, he's got chores on Saturday morning." Your message expresses your need to see the child's potential fully realized.

4. You are interviewing a rape victim when her boyfriend barges into the room and demands, "What's going on in here?" Your message expresses your need to continue the interview.

(Note: In such situations, never ask the woman, *in front of the man*, if it is all right for the man to stay during the interview. Lead the man outside when you give him your message. Later, when you and the woman are alone, you can ask her whether she would like to have him present, but always make it appear that the decision to have him wait outside is *yours*. It is possible that she is afraid of him and will feel compelled to agree to his staying if she is asked about it while he is in the room. If she is fearful or embarrassed, the quality of the interview will be compromised.)

5. An elderly woman is trying to decide what to do about her need for help. The decision is between staying in her own home with assistance, or selling her home and entering a nursing home. She is very torn. You have arranged for help, which seems to be working well, and you visit her each week. During your visits, the woman discusses with you her options. The decision is a difficult one for her. When you visit her, a woman who lives next door invariably appears and offers her advice and expresses her doubts. Your message to the neighbor expresses your feeling that her behavior is not helpful.

✗ Chapter 12

Addressing and Disarming Anger

INTRODUCTION

People do become angry. They express anger and hostility in ways we might find quite unpleasant. We can expect that there will be times when the people with whom we are working will forcefully express their anger. As professionals, it is helpful to view the anger as a clue to other underlying issues or as a clue to problems that need to be resolved. Using the anger to help us better understand the other person is better than reacting to it defensively or personally. If you are an effective, empathic listener, you will hear these underlying causes and respond in a manner that disarms rather than provokes the anger.

COMMON REASONS FOR ANGER

When clients are angry, it is often because of one of the common reasons listed here.

    1. The client is angry about something the agency has done. The agency in which you work will have policies and regulations that you must follow. Sometimes the agency is bound by state and federal laws. These laws can work better for some clients than for others. Clients who feel that the agency's policies have caused them to be treated unfairly or with insensitivity to their particular circumstances may react angrily.

    2. The client is angry about something you have said or done. As we noted earlier, there will be times when the client or the client's friends and relations will have a problem with something you have said or done. Without your intending that it should happen, a client may completely misunderstand what you have said or may misread your intentions. On the other hand, you may not always be completely tuned in to where the client is at any particular moment and may unwittingly say or do something the client finds upsetting.

    3. The client is fearful. Many clients are frightened by the turn their lives have taken. The changes that have occurred that brought them to your agency may make them feel as though their lives are out of control. They may attempt to reassert control through the use of anger, and they may lash out at you because you are the safest target or the closest person at the moment.

4. *The client is exhausted.* Some clients you see will be exhausted. These people may have been grappling alone with an issue or problem for a long time, or the circumstances they are facing now may be taking all their energy. They sense that they may not be able to carry on, which may cause them to direct anger at you.

5. *The client feels overwhelmed.* Other clients feel overwhelmed by problems. They may feel as if they cannot handle all that is facing them. Sometimes they feel the extent of the burden is unfair, and so they lash out at you.

6. *The client is confused.* Some people are confused by policies, circumstances, others' reactions to them, or the steps they must take to right a difficulty. Rather than admit to feeling confused, it is better for some to become angry and blame the system or you or your agency.

7. *The client feels a need for attention.* Some people feel insignificant and demeaned. It may have nothing to do with you, and it may very much relate to a lifetime of living in the margins or having one's problems or contributions trivialized. These people need to feel valued and worthwhile. The problem for you is that your best efforts may not always be good enough. Sometimes people are extremely tuned in to slights and suspected rejection. They may become angry with you for reasons that you feel are unfair or unwarranted. As always, you are the professional person and need to speak to the condition of the client in a professional manner.

There are many reasons why a person becomes angry. Knowing how to disarm anger is important. It will allow you to move toward a more meaningful dialogue and a better resolution.

WHY DISARMING ANGER IS IMPORTANT

You cannot be as effective in your work if you are dealing with a client who is angry. The client cannot be expected to move the relationship to another level; but you, as the professional, can be expected to practice the techniques that will allow the relationship to move beyond the anger.

The major reasons for disarming anger are as follows.

1. *Disarming anger diffuses the anger, making it less of an obstacle to true understanding.* People who are angry cannot really hear each other. If you are genuinely interested in why the client is reacting in this manner, you need to reduce the anger so that you can better understand what is fueling these strong emotions.

2. *Disarming anger shows the other person that you respect the message even if the way it is expressed is not helpful.* By moving to another level beyond the anger, you can indicate to angry clients that their concerns are important to you even when you are having trouble with the way they are addressing these concerns.

3. *Disarming anger allows you to become aware of the actual problem.* Only when you have disarmed the anger can you and the client actually address the underlying concern. As the client feels understood or heard, he or she is more likely to begin to collaborate with you in looking at the problem and the solutions.

4. *Disarming anger allows you to practice empathy, seeing the situation as the other person is seeing it.* Disarming anger is an important part of establishing rapport. If you become angry yourself, you are caught up in your own feelings and needs at the moment. On the other hand, if you think about the reason the person is angry and you speak to that situation or to those feelings, you are responding empathically. This lets the client understand that you are not going to engage in an angry exchange, but you are going to respect the client's concerns and feelings.

5. *Disarming anger focuses on solving the issues and problems, and not on who is to blame.* Disarming anger techniques do not allow for exchanges of blame. Angry clients may hope for such an exchange with you wherein they blame you and you defend yourself, often by blaming them in return. The purpose of disarming anger is to fix those things that legitimately need to be fixed.

Many people sound angrier than they mean to. They are often anticipating the angry response of the other person. As human service workers, we read anger as a signal that the client's needs have not been met, and we focus on resolution of the problem that has caused the angry emotions, regardless of whether we think the client's anger is legitimate or not.

AVOIDING THE NUMBER ONE MISTAKE

Countless times human service workers encounter people who are openly angry. Many of those workers choose to take that anger personally. Taking anger personally is the number one mistake when dealing with an angry person. It is a foolish mistake to make.

As noted earlier, people are angry for a number of reasons. Some of these reasons have nothing to do with the worker specifically. Other times the anger may be caused by something the worker or the agency has done, and the anger may be rude and denigrating. Nevertheless, beyond disarming the anger, it is important that when you encounter an angry client, you refrain from taking the anger personally. This is demonstrated in the following example of a worker who chose to take the anger of a client personally.

Client	Worker
— Where the hell were *you* on Tuesday?	— What do you mean?
— Where the hell were you? I came in to get a voucher for food, and you weren't here.	— Why are you shouting at me? I wasn't here, but you don't have to shout.
— I do have to shout! You say to come in here for a voucher, and I did that, and you were not even here. Where the hell were you?	— Look, Mr Peters, I don't have to tell you where I was. If you came in and I wasn't here, why didn't you tell someone else what you needed? I'm not the only person who can help you.
— I get so damn tired of the way you guys act like prima donnas. Who the hell gave you the right to tell all of us when to come and when to go? You say come in, I come in, like a fool, and *you* decide you'll just go someplace else.	— Well, if that's the way you feel, you certainly don't need *my* help. I've spent quite a lot of time with you, may I remind you? You have gotten a lot from this agency. I'm not sure I'm going to put up with this shouting at me.
— Well, what are you going to do about it? I can tell you that you are a piss-poor caseworker if I want to. I can't do much else around here, but I *can* do that!	— You're an idiot. Go out and get the voucher from Mrs. Charles, bring it back here, and I'll sign it (begins reading papers on her desk).

In this example, the relationship is damaged, and there is an unsatisfactory resolution. Bitter feelings remain for both the worker and the client. There is a better way to handle situations like this one. The next section explains how to use a four-step process to deal with anger.

THE FOUR-STEP PROCESS

In his book *Feeling Good,* David Burns (1984) suggests a four-step process for disarming anger. The material in this section is adapted from his book. First, we will look at the individual steps, and then we will look at how these steps work in actual practice.

1. Be appreciative. It is frightening enough to tell a person you are angry about something he or she said or did. You will put the client at ease if you can say something like "I appreciate your coming to me with this" or "It is helpful to know how you are feeling about this" or "Thank you for bringing this up."

2. Ask for more information. If the client is upset, he or she may be skimming the surface of the issue. In order to understand the issue better, ask for particulars. Do not grill the other person or sound defensive. You might say, "Can you tell me when this happened?" or "I guess I'm not sure when this happened. Can you help me out?" or "How often did this happen?"

3. Find something with which you agree. Never make up something just to sound agreeable, but see if there is not some little piece of what the client has said with which you can agree. You might say, "I think we have probably done this in the past" or "I can see where you would feel that way" or "There probably was a lapse the day you are referring to."

4. Begin to focus on a solution. Focusing on a solution should involve *collaboration* whenever possible. Remember, however, that the client owns the problem; the extent of your assistance is a conscious and strategic decision that you must make.

Begin with your objective point of view. Listen to the client's point of view. Then decide if you will make changes or leave things as they are. You might, after explaining your point of view, say, "I can see where you thought that. I think from now on I'd like to write you ahead of time" or "I can see where you thought that. Right now we are really short-staffed so writing to you ahead of time isn't really an option. I'm glad you brought this up. In the future, we will take another look at it." Own your perceptions and own your decision regarding the problem. Use "I," not "you."

Let us return to the situation we looked at earlier and see a more effective approach for handling the problem.

Client	Worker
— Where the hell were *you* on Tuesday?	— I'm not sure I know what you mean.
— Where the hell were you? I came in to get a voucher for food, and you weren't here.	— What time were you here on Tuesday?
— Oh, about 2:00. You say to come in here for a voucher, and I did that, and you were not even here. Where the hell were you?	— (Using active listening first) Well, this must have been really inconvenient for you. I appreciate your telling me about this. Did you ask anyone else to help you?
— No, I didn't. I didn't know I could.	— Sure you can. I can see where you would think I didn't care about your getting food when you expected me to be here.

— I know. We didn't have any dinner Tuesday night. Just potato chips and cheese—oh, and there was a little milk left.

— Thanks.

— I'm really sorry that happened. Let me see that you get the food you need today, and let me explain better than I did the other day how this works. If I or any of the other workers tell you to come in for a voucher, then you can come right to the office, and whoever is doing the intakes can take your information and see that you get the food you need. This shouldn't have happened this way, and we don't want it to happen again.

All the elements are present in this exchange. Rarely, as is true of this exchange, are these elements present in any precise order. In the example, the worker asks for more information with genuine interest (What time were you here on Tuesday? Did you ask anyone else to help you?). She goes on to express appreciation (I appreciate your telling me about this). She indicates that she agrees with the way the client viewed the situation (I can see where you would think I didn't care about your getting food when you expected me to be here). Finally, she moves on to focus on a solution (Let me see that you get the food you need today, and let me explain better than I did the other day how this works).

The worker in this example does some other things that make it clear she is not going to take the client's anger personally. She uses active listening (Well, this must have been really inconvenient for you), letting the client know that he is being heard and respected. She also takes some responsibility for the mix-up (and let me explain better than I did the other day how this works).

We might change this vignette just a bit. Perhaps the worker actually did explain to the client on the phone before he came in how the agency works. There are many reasons why he might not have heard her: anxiety over trying to make sure his kids would eat that night, anger over having to go to the agency in the first place, uncomfortable feelings of helplessness or inadequacy over his inability to fix his situation on his own, and the stress of not eating and having hungry children at home.

Although the worker may not know specifically what has generated the angry outburst, she is fully aware that there are forces at play in this man's life beyond his need for her to be present when he arrives at the agency. For that reason, she remains respectful throughout the entire exchange, and she moves with genuine interest and concern through the steps of disarming anger. In other words, she does not take his anger personally and feel a need to confront it with anger of her own.

WHAT YOU DO NOT WANT TO DO

Do Not Become Defensive

Do not fall into the trap of defending yourself. It is okay to have made a mistake or to be wrong. If you begin to defend yourself, it makes the other person angrier, and you lose an opportunity to really resolve the problem. This is demonstrated in the following responses to a person who feels the worker did not spend enough time with her:

Correct	Incorrect
— I might have cut the interview short.	— I'm doing the best I can. I certainly gave you the time you needed!

Do Not Become Sarcastic or Facetious

When you thank the other person for their comments or agree with something they told you, it is possible that you will sound sarcastic or facetious. This is especially true if you are feeling defensive, as illustrated in the possible responses to a man who works and is frustrated because he needs a later appointment but keeps getting an early morning appointment:

Correct	Incorrect
— I'm glad you brought this up again. We really do need to get this straight.	— Here we go again! Thanks for telling us, again, how inefficient we are.

Do Not Be Superior

It is all right for you to be wrong in your perceptions or behavior, and it is all right for the client to be wrong too. If you feel especially threatened or angry at your clients, it may be tempting to denigrate them in some way, pointing out how little they actually know about the situation or how little experience they have and how much more knowledgeable you are. This is the case in the following incorrect response to a woman who thinks her daughter should have different services:

Correct	Incorrect
— We ought to look at this more closely. I'm glad you told me about this. You may be right.	— The services have been chosen for your daughter by professionals in the field of child development, and they know what it is she needs.

Do Not Grill the Client

In order to better understand the problem from the client's point of view, you will need to ask questions. Avoid grilling the client by asking numerous questions, one after the other, in a doubtful tone of voice. If the client is nervous, you will only make his or her nervousness worse. Most people grill others in a triumphant attempt to prove the other person wrong. That is not your goal here. Your goal is to genuinely try to understand. Examine the following responses to a man who believes his aunt is being neglected by the agency:

Correct	Incorrect
— Tell me more about what you see happening with us and your aunt. We may need to look at this situation more closely.	— When exactly did we fail to come out to your aunt's house? How often did this take place, and did she ever tell us about this before? We need to know specifics before we can determine if this is a real problem. What other problems did you encounter with us?

LOOK FOR USEFUL INFORMATION

You can benefit from the feedback you are receiving if you really hear it. Sometimes the client is bringing you valuable information that will help you to make constructive changes in yourself or in your agency.

In one agency, there were a lot of angry clients calling for help. They all had been discharged from a certain program without follow-up services or with follow-up services that had not been confirmed. The agency was grateful for the clients' feedback and developed a questionnaire for the receptionist to use when such calls came in. Gradually, a picture emerged of precisely what was wrong and how to fix it. In this example, an entire agency benefited from the clients' feedback. A more efficient operation will keep clients from returning with recurring problems and will save money and time for other clients.

MANAGING AN ANGRY OUTBURST

On rare occasions, people become so angry they seem to be about to lose control. Their anger moves from rational expressions of anger to increased belligerence, threats to the safety of others, or actual aggression toward people and objects in the vicinity. Research shows that staff persons play an important role in defusing these explosive situations. An even tone of voice, continued active listening, and relaxed movement will work best.

Lisa, a nurse in a community program for the mentally ill, discovered Phil eating lunch one day with a gun lying by his plate. He had been angry about his medications earlier in the day, but that problem seemed to have been resolved. Instead of quietly approaching Phil and suggesting the gun be left with the nurse until the end of the day, Lisa became hysterical. Rushing about the room, she loudly began to clear out the startled patients, thrusting them through the door. "Call the police, call the police," she kept shouting to other workers. Phil, alarmed by her actions, grabbed his gun and pointed it at her. He began to yell at Lisa, "Just shut up, shut up, before I shoot you. Be quiet." Lisa dashed from the room and cleared the entire building. Police came from every direction. The area was cordoned off, and a stand-off ensued into the afternoon.

Lisa's loud, hysterical tone of voice, her panic, and her hurried actions all combined to make Phil agitated. Before long, the situation had escalated. What Lisa should have tried first was asking Phil to come with her to another part of the building. If Phil left the gun at his place, another worker could have secured it. If he brought the gun with him, Lisa might have said matter-of-factly that perhaps it would be better to leave the gun with the staff for the time being. If Phil gave her the gun, she could have taken steps to secure it.

Even if Phil were resistant and wanted to keep his gun, the staff could have asked the clients to bring their lunches into the group rooms for after-lunch groups. If they made this request in a tone of voice that indicated that it was nothing out of the ordinary, clients would have complied. In the meantime, police could have been called to come quietly and help to disarm Phil.

In another situation, Jim, a young mental health case manager, was working with Alex. Alex wanted to go into the hospital, fearing that he was getting sick again and would hurt someone. On that particular day, there were no beds, and Jim's supervisor suggested he help Alex find an alternative to hospitalization until a bed was available. Jim was afraid of Alex, and each time Jim explained that no beds were available and that an alternative needed to be found, Alex grew more belligerent.

The last time Jim returned to report that there were still no available beds, he did so in what he thought was a very firm manner. Reasoning that Alex seemed about to become uncontrollable, Jim assumed that if he approached him with a firm, superior tone of voice, he could keep Alex from getting any angrier. In fact, Jim's superior tone was heard by Alex as denigrating. He began to break furniture to demonstrate to Jim "just who is in charge here."

Remaining matter-of-fact, practicing empathic active listening, and remaining calm are important in maintaining a controlled situation. If you or your patients are in danger, certainly the first thing to do is to secure the safety of everyone. If, however, situations escalate because the workers fuel the situation, inflaming the client's anger by becoming loud, agitated, or angry themselves, the situation can rapidly spiral out of control. Remaining calm and moving deliberately to prevent a dangerous situation from worsening is the responsibility of the professional.

INITIAL RESPONSES TO ANGER

Instructions: In the examples that follow, formulate an initial response to the anger and criticism you hear. Look at the steps for disarming anger, and use those that seem appropriate. The four steps are: (1) thank the person for his/her comments, (2) ask for more information, (3) find some point on which you can agree, and (4) begin to look for a negotiated solution.

1. A man is coming to your agency for assistance after a violent crime. He wants you to do more for him than you think is wise. You have been very helpful in ways you could, but you have also insisted he do some things for himself because you do not want him to become dependent. He is frustrated, and one day when you suggest to him that he try to call his lawyer himself, he blows up and yells, "This crazy place, sucking up the taxpayers' money—and for what? I get so sick and tired of your trying to make me do everything when that's what you're paid for. A bunch of idiots is what you are! Incompetents! Sure, I can do it myself! If I wanted to do it myself, I wouldn't have come to you, would I?" What is your initial response?

2. A woman who is in your shelter feels neglected. Twice you are interrupted when you are talking to her because of severe emergencies. You apologize both times and continue your discussion with her; but you are short-staffed, and things at the agency are unpredictable. The second time this happens, when you are able to get back to her, she cannot remember what she had been saying. That upsets her. She says, "You all sure can find plenty of reasons to avoid talking to me. Every time I sit down to talk about my case, you get up and run off. Now I can't remember

where we were. I don't see you running off when you talk to Alice or Cindy. Just seems like every time I need help, well, you have something more important to do." What is your initial response?

PRACTICING DISARMING

Instructions: Following are some opening sentences said by angry clients. It is up to you to develop the exchange, including more information regarding what the client is angry about and the responses of the worker. You do not have to use the disarming steps in any particular order. See if you can add some active listening and open questions as you go along. See if you can put yourself in the clients' shoes and empathize with their feelings.

Client	Worker
— (Talking loudly, and banging his fist on the desk) I have a beef to pick with you! Little jerks run this agency, a bunch of little jerks! You tell me I'm a mental case and then give me medicine that makes me feel like a nut case.	

Client	**Worker**
— (Barging past the receptionist to the worker's office, obviously angry) My kids and I are hungry! Know what that means—to be hungry? We're hungry and you . . .	

Client	Worker
— (Sitting down hard in the chair and glaring at the worker) Forty billion times I say, "Look, I work every day, ev-er-y sin-gle day." You just bop along, giving any old appointment that suits *you*!	

✗ Chapter 13

Putting It All Together

INTRODUCTION

You are now at the point where you can practice holding entire conversations with clients using all the skills you have learned. At first, this will seem awkward and mechanical to you. You will hesitate as you try to decide whether to use an active listening response or an "I message." You will feel that you are engaging in a dialogue that is halting and unsatisfactory. With practice, the skills become second nature. Your responses will be smooth and sound unrehearsed and genuine. The following exercises are designed to help you begin to put all the skills together in the same conversation.

PUTTING IT ALL TOGETHER

Exercise I

Instructions: Follow the instructions for each question based on the communication skills you have learned in class.

1. You have been called to an elderly woman's home. The family is upset that she is refusing to leave the home for an evaluation at the hospital even though she can no longer walk, is incontinent, and cannot get to the kitchen to fix meals for herself. She lives alone. When you arrive, you find her daughter and son-in-law exasperated from their attempts to convince her to go to the hospital for a medical workup, as her doctor has recommended. You enter the room, and after introducing yourself and finding a place to sit down near her, you ask her an open question. Write your open question.

2. The woman tells you why she is not going to the hospital. "I don't want to," she says, "Why should I? I'm an old lady. I've lived a full life. I want to die right here. Harry [her husband] died down at the hospital. He thought he was coming home, and he never did. He didn't want to go down there, and we all made him go. We thought he would get well, come back home. He never did. I'm certainly not going through that." Write a word that describes the feeling you think the woman is expressing, then write your empathic response.

   FEELING:

   EMPATHIC RESPONSE:

3. You are quite concerned about this woman being at home alone. As you sit there, you can see how extremely frail and weak she is. After awhile, you express a very gentle "I message" or concern that *you* feel regarding her being home alone. Write your comment here.

4. In response, the woman answers, "You know, I never said I wouldn't go anywhere! I just said I wouldn't go to that godforsaken hospital where Harry died." Write an opening sentence that shows you are beginning to collaborate on a solution that might work for her and still accomplish what is needed.

PUTTING IT ALL TOGETHER
Exercise II

Instructions: Follow the instructions for each question based on the communication skills you have learned in class.

1. You are seeing a 40-year-old man for the first time. He has a diagnosis of schizophrenia. This is a chronic condition that started when he was 19 and went away to college. Today he has come in to see if you can help him with a better place to live. He tells you he just wants to "live somewhere else, that's all." You sense that he is upset. You ask him an open question. Write your open question here.

2. In response, the man looks away, and then slowly begins to talk about the kids in the neighborhood. He says they make fun of him and try to stop him when he walks home at night. "I get sort of sick when I see them. The other night I stayed out in the alley most of the night. It just seems like they are out to get me. They throw bottles at me and call me 'psycho.' I don't know. I just stayed out so they wouldn't get me." He looks sad and embarrassed. Write a word that describes the feeling you think the man is expressing, then write your empathic response.

   FEELING:

   EMPATHIC RESPONSE:

3. You are quite concerned about the stress of this situation for him, knowing stress can trigger a relapse. You are not clear what else he does during the day or what friends and family he might have for support. You ask him an open question about his activities. Write that question here.

4. The man tells you that he is enrolled in Goodwill during the day and that he has been taking the bus on his own to the center where he works in the retail shop.

"When I have to take the bus, I stand on the corner, and they always see me. Or when I get home and get off the bus, they seem like they are waiting for me." Write a word that describes the feeling you think the man is expressing, then write your empathic response.

FEELING:

EMPATHIC RESPONSE:

5. You still want to determine if this man has any supports other than Goodwill. Write the open question you will ask him.

6. In response, the man answers, "My family got funny on me years ago. I have a sister. She lives in Lemoyne, and she always has me come over at Christmas. She is the only one that bothers with me. The family always acts nice at first, but I can tell I get on their nerves. I don't stay real long." Write a word that describes the feeling you think the man is expressing, then write your empathic response.

FEELING:

EMPATHIC RESPONSE:

7. The man continues, "But I have this friend from Goodwill—Paul. He told me I could come to his house if things get bad. He didn't want me to stay in the alley like I did." Write a response to the content of this statement.

8. The man goes on to say, "Yeah, Paul is a real good friend. The other day when I ran out of Mellaril, he lent me some of his until I got my prescription filled down at the county." You are concerned about his use of another patient's prescription. Write your "I message" to express your concern here.

9. The man says, "Well Cindy, down at Paul's apartment building, she works for the county, and she got upset too, like you. She said Paul ran out of his stuff too soon to get a refill, and they had to see the doctor especially to get him more pills." Respond here to the content of his comment.

10. Now that you know what supports are in place, you need to know what he thinks about finding another place to live. Ask the man an open question to learn what ideas he has about finding another place and to begin collaborating on a solution. Write the open question here.

11. The man answers by saying, "Well, I was thinking I could get into the program Paul is in. It's part of Goodwill. They told me I was doing good and didn't need to be in that apartment building, but I know a lot of the people there, and I could, well, feel better if I lived there—or down near there somewhere." He is referring to a personal care boarding home that is sponsored by Goodwill. Write a statement that indicates you want to collaborate with him on the issue of a new place to live.

PUTTING IT ALL TOGETHER
Exercise III

Instructions: Follow the instructions for each question based on the communication skills you have learned in class.

1. Mrs Sylvestri is worried about her son, Manuel. She tells you that he seems to have trouble concentrating in school and that the teachers have told her he is "a daydreamer." He is falling further and further behind, and she is wondering if her son is reacting to the recent death of his father in a construction accident. She says, "I don't know if you saw it on TV or not. He was working on that sewer project down in Carsonville, and the walls of the trench fell in on him. All his buddies were there, trying to dig him out. It was on the news." You respond with active listening. Write a word that describes the feeling you think the woman is expressing, then write your empathic response.

   FEELING:

   EMPATHIC RESPONSE:

2. Mrs Sylvestri seems grateful for your understanding and goes on to describe the fact that Manuel saw the newscast of the accident at a friend's house. "He was over there playing about the time the news came on," she said, "I hadn't gone to get him because I had so much to take care of, the hospital and then the funeral home and calling his brother to decide what we were going to do. And Manny saw it on the news. The neighbors didn't know, not yet." Ask an open question about her husband.

3. She tells you he was a good father and a good provider, that Manuel is their only son, and that they were unable to have other children. They came together from Puerto Rico nine years ago and worked hard to buy a small house in Carsonville and serve the community. "Now he is gone, just when Manny is about a teenager, and . . ." Give an active listening response. Write a word that describes the feeling you think the woman is expressing, then write your empathic response.

   FEELING:

   EMPATHIC RESPONSE:

4. Mrs Sylvestri asks you if you can help with Manny. She tells you what a "good kid" he is and how much he helps around the house. "He has cousins, all his uncles and aunts, they live in Carsonville or around, so he has family, and me, but he won't pay attention in school. He hurts." Ask her an open question that will help you to know what she might have in mind for Manuel.

5. Mrs Sylvestri responds with the fact that the school wants him tested for attention deficit disorder, but she does not think he is "mental or anything." Instead, she believes he is grieving and says, "I don't know what you do for a kid who just lost his father like Manny did. I don't know." Using an "I message" suggest the children's Arbor House that specializes in children and grief.

6. She looks interested. "Are there other kids, that many other kids who have problems like Manny? So sad to think. What do they do there?" You describe the group sessions and the activities they have for the children and the age range of children Arbor House accepts. Then you move toward collaborating with her on a way to engage in these services. Write your comments here.

PUTTING IT ALL TOGETHER
Exercise IV

Instructions: Try this exercise as often as you need to in order to become more proficient in the communication skills. Find a partner with whom to practice the skills. Using a tape recorder, tape your conversations.

1. First, you be the worker and let your partner play the part of a person who is seeking help for a personal problem or condition. Play the tape back. Critique your efforts with your partner, looking at ways you might improve your responses.

2. Ask your partner at the end of the dialogue how your responses felt to him or her. Did he or she feel reassured, understood, supported? Did your partner feel genuine interest and concern on your part?

3. Next, reverse roles, letting your partner be the worker while you take the role of the client. Again, tape the conversation and go over the tape together, looking for ways to improve your partner's skills.

4. Give your partner feedback about how his or her responses felt to you.

5. Once you have made a good tape of you as the worker responding to the client, hand it in to your instructor or play it for the class. Be prepared to receive additional pointers for improvement. At this stage in your training, every word can be examined. Soon, however, you will begin to know exactly what to say.

Meeting Clients and Assessing Their Strengths and Needs

X Chapter 14

Documenting
Initial Inquiries

INTRODUCTION

In many agencies, phone inquiries are logged on the computer and kept on file there. The New Referral or Inquiry form used in this text contains the kind of information the client would be asked for during an initial call to the agency, particularly if the person were asking for an appointment. Learning to gather this information properly is important to ensure that the person's situation is handled well from the beginning.

This first form, therefore, is used to take information from people calling *on the phone* to obtain services. You are to ascertain what the problem is, in brief, and set up an intake appointment where the caller will be seen in person for a more in-depth history and evaluation. Remember to use all the communication skills you have learned in previous chapters to make the client feel at ease while describing the problem. Asking for help is difficult, but it is less so if the phone worker is empathic and accepting.

STEPS FOR FILLING OUT THE FORM

A typical form for phone inquiries is found in your appendix. It is titled "New Referral or Inquiry." The following is a step-by-step process for filling out this form. Take the blank form from the Appendix in the back of your book, make copies of it, and follow the steps provided here:

1. Place the client's name, sex, date of birth, and address at the top of the form.
2. Place a home phone number on the form, and the work number of the client if the client is working.
3. The person will either:

   a. be a minor and have a parent or guardian, in which case you circle or underline "parent" on the form and write in the name of the parent, or

   b. be an adult with a spouse, in which case you underline or circle "spouse" on the form and write in the name of the spouse, or

   c. be neither of these, in which case you write N/A in big letters on that line

4. If the person is employed, place the name of the employer on that line. If the person is not employed, place N/A on that line.

5. If the person is in school, place the name of the school (complete with what kind of school—college, elementary school, high school) on that line. If the person is not in school, place N/A on that line.

6. The client will either be:

   a. a self-referral, meaning the person found out about your agency through the phone book or a friend and called in on his or her own. If that is the case, write "self" on that line. Most calls are self-referrals, or

   b. referred by a doctor or other professional. In that case, place that person's name on the line. You are asking the client, "Who referred you to our services?" Answers might be Dr. Graham Smith or Attorney William Burns.

Please Note

You cannot accept a case and fill out this entire form on the word of another person. The other person can refer the client to your agency, but the client must call in order for the intake to be valid. There are these exceptions:

If the client is incapable of calling due to a mental health emergency. That situation requires other forms and does not apply here.

If the client is a child, the parent or guardian can call for that child.

If the client is a very infirm or frail older person, a family member or friend can call for that person.

---

7. Under the section marked "Chief Complaint," always tell why the person called *today*. Do not say the person called today because her husband is abusing her. The husband may be abusing her, but what made her go to the phone today? Here are some reasons a person might give for calling today:

- Today the person decided he or she cannot go on.
- This morning the employer insisted the person get help.
- She saw a medical doctor within the last 48 hours who told her she needs counseling.
- He just had a fight with his spouse.
- He just hit his child.
- She thinks she will hit her child and quickly called you to stop herself.

In filling out the "Chief Complaint" section, capture why the person called on this date and not on some other day. Begin this section with "Client (or the person's name) called today because . . ."

CAPTURING THE HIGHLIGHTS OF THE CHIEF COMPLAINT

You have a small space and can use very few sentences to describe why the client called today and not some other day. In thinking about what the client has told you about why he or she called, choose the most important points. Here are some examples:

John Haulik called today because his employer requested he seek help for drug problem (crack). In last two weeks, he has missed or been late to work every day. Sleeping on the job and cited for safety violations. Client sounded distressed and eager to begin treatment.

Jane Wilson called today after a serious fight with her husband involving physical abuse. Jane states husband has been verbally abusive in past, but not physically. Client is hospitalized and looking for alternative safe living arrangements upon discharge. Client sounds depressed and anxious, but cooperative.

The following are some guidelines to keep in mind:

1. *Keep the reasons from being too complicated.* Do not make these first cases psychiatric emergency situations that need either immediate attention or a commitment. In other words, in this first exercise, do not create clients who are hearing voices, are contemplating suicide, have made a suicide attempt, or are a danger to others.

2. *Be very specific.* Do not use general descriptions such as "her husband beats her" or "she lives with an alcoholic" or " he has been having a hard time at work." Tell when the last beating was, what the most recent problem with the alcoholic husband was, and what the most recent problem was for the client at work.

3. *Keep the reason for the call brief.* Do not include a lot of background information, as that will be acquired when you do the social history at the time of the evaluation. Give just enough background information to let the next worker know the context of the client's problem. For example:

> Client called today because she was severely beaten by her husband on Tuesday during an argument over dinner. Client was hospitalized and is seeking alternative shelter. There is a history of domestic violence, which the client feels has worsened in the last nine months.

> Client called today because his employer warned client that without evidence of treatment for problems with alcohol, employment may be terminated or suspended. Client admits to drinking while on the job today. There is a pattern of binge drinking followed by missed work and problems with coworkers.

EVALUATE THE CLIENT'S MOTIVATION AND MOOD

Complete your note with a single sentence that indicates how the client seemed to you. For example, you can mention how the client sounded. Did the client seem depressed, glad to have reached you, relieved to be getting help, guilty over what has happened? Did the person seem eager to engage in services, skeptical that you can help, cynical about complying with forced treatment, or cooperative? Here are some examples of sentences that might summarize how the client seemed to the phone worker:

- Client expressed a desire to begin treatment immediately.
- Client seemed depressed by the circumstances, but motivated to follow through with services.
- Client expressed skepticism that anyone could help him, but seemed motivated to seek help.
- Client was tearful and seemed depressed during the interview.
- Client seemed agitated by these recent developments and somewhat unwilling to follow agency procedure.

COMPLETE THE FORM

The following steps complete the process described in the previous section entitled "Steps for Filling Out the Form."

8. Under "Previous Treatment," keep the notes brief—just note when, where (and with whom if you know that), and for what. Keep from being too wordy in this section.

Correct

Seen in June of 1989 for six months by Dr. Piper, Waldenham Clinic, for postpartum depression.

Incorrect

Susan saw Dr. Piper at the Waldenham Clinic, 432 Muench Street. She started to see him in June of 1989 after her first son was born and continued to see him for six months. He was treating her for postpartum depression.

9. The intake is "taken by" you. *This is the first place your name is to appear on this form!* Put the date of the intake next to your name.
10. Under "Disposition," note the name of the person to whom you refer the new client for intake and the date of the intake appointment. In many settings, the person who handles the phone inquiries is not the same person who sees the clients when they come in for their first appointments. For training purposes, we will assume that you will be doing both the phone inquiry and the client intake, in which case you would write your own name, along with the date of the intake appointment, on that line.
11. Under "Verification Sent," write "Yes" and the date. The date you use here is the date you send out the form, usually the same day on which you take the phone inquiry.

Figure 14.1 shows an inquiry form filled out correctly. Look at the form to see how the worker filled in each element.

After a person has inquired about services from your agency, it is important to bring the person in for a more thorough history and evaluation of the problem, if the person is seeking services. You will set up an appointment for the caller on the phone at the time of the call or soon after you hang up. The next step is to send a letter verifying or confirming this appointment.

SENDING A VERIFICATION

Note: Forms referred to in this section are in the Appendix.

The purpose of the verification letter is to confirm for clients the appointments that were made with them for an initial intake in the office. In this way, the agency hopes to cut down on the number of missed appointments and the number of hours reserved for a client that are not used by the client.

Today many agencies have more clients than they can see easily. Long waiting lists are the result. An agency cannot afford to waste an hour on a client who does not come in for a scheduled appointment. Although it may give the individual worker a much-needed break and time to catch up on paperwork, it is an hour for which the agency will not be reimbursed because no services are given. For that reason, most agencies send out a verification letter to remind clients of the appointments that are reserved for them.

NEW REFERRAL OR INQUIRY

CLIENT _Karen Elaine Markley_ SEX _F_ DOB _4/9/69_

ADDRESS _1234 Pleasant St._

Anytown, PA ZIP _01234_

HME TELEPHONE _555-555-5555_ WK TELEPHONE _555-555-5555_

PARENT OR SPOUSE _John H. Markley_

EMPLOYER _Evansville Township_

SCHOOL _NA_

REFERRED BY _Dr. Walter E. Carmichael_

CHIEF COMPLAINT AND/OR DESCRIPTION OF PROBLEM _Client called today because she_ _suffered an incapacitating anxiety attack at work. A coworker took her to Dr. Carmichael who referred_ _client. She states that she suffered first attack 4 years ago following birth of son. These attacks are_ _increasing in frequency, particularly at work. Client sounded distressed and was tearful at times. She_ _seems eager to begin treatment._

PREVIOUS EVALUATION, SERVICES, OR TREATMENT _Was seen by Dr. Allen Peters, Winston_ _Clinic, from 6/96 to 7/97 for general anxiety and agoraphobia_

TAKEN BY _Marcia Andrews_ DATE _9/13/98_

DISPOSITION _Referred to Marcia Andrews for intake 9/21/98_

VERIFICATION SENT _Yes (9/13/98)_

Figure 14.1 Sample of new referral or inquiry

Following is a step-by-step procedure for filling out the verification letter. A blank verification letter is found in the Appendix at the back of the book. Make copies of that form and use that to follow the steps in filling out the letter.

1. On the Verification form, be sure that the date you send it out is the same date you said you sent it out on your New Referral or Inquiry form.
2. Be sure to address the client by name.
3. Fill in the date, time, staff, and location of the interview. The date is the date you listed under "disposition" on the inquiry form. You can decide on a time. The staff person will be you. The interview will take place at the Wildwood Center.
4. Sign your name. Your signature should line up precisely under "Sincerely" and over "Case Manager." Do not sign out to the right.

Figure 14.2 contains a sample verification letter with all the information added.

The next step in the process is when the person actually comes to the agency for a more thorough evaluation of his or her situation. This is sometimes called an intake

Wildwood Case Management Unit

Date _____*9/13/98*_____

Dear *Mrs. Markley:*

This letter is to inform or remind you that you have an appointment scheduled:

Date: _____*9/21/98*_____

Time: _____*2:30 P.M.*_____

Staff: _____*Marcia Andrews*_____

Location: _____*Wildwood Center*_____

Please contact me if you have any questions or if you need to reschedule.

Sincerely,

Marcia Andrews

Case Manager

Figure 14.2 Sample verification of appointment

appointment. In this chapter, we have practiced phone intakes; in the next chapter, we will turn to the first appointment with the client and examine how to prepare to meet the client for the first time.

INTAKE EXERCISES

Exercise One   Intake of a Middle-Aged Adult

Instructions: Using the blank form in the Appendix entitled "New Referral or Inquiry," develop a client intake. You may use any problem that would ordinarily come to the attention of a social service agency. Your client should be an adult, at this point, calling on his or her own to seek services.

In developing your client and your client's problem, read the American Psychiatric Association's *Diagnostic and Statistical Manual of Mental Disorders (DSM-IV)** and look at books

---

*Note that *DSM-IV* was revised in 2001 to become *DSM-IV-TR*.

and articles on specific problems (such as domestic violence, alcoholism, divorce, and depression). Look at the chapters in the companion textbook, *Fundamentals for Practice with High Risk Populations,* for information on how your client might be feeling and what issues or problems the client might be facing when he or she makes the first call to your agency.

If time permits, do several adult intakes with each client having a different reason for calling.

Exercise Two   Intake of a Child

Instructions: Using the blank form in the Appendix entitled "New Referral or Inquiry," develop a client who is a child, a person under 16 years of age. In this case, a parent or guardian would be calling on behalf of the child. A doctor, a school counselor, or a teacher may have referred the parents to you, or the parents may have felt they needed help and sought your services without a referral. Typical problems confronting children are problems with school, behavioral problems, and adjustment problems to events like divorce or the death of a parent.

In developing your client and your client's problem, read the *DSM-IV* and look at books and articles on specific problems common to children. Look at chapters (particularly those on children's mental health and on mental retardation) in the companion textbook, *Fundamentals for Practice with High Risk Populations,* for information on how your client and your clients' parent might be feeling and what issues or problems the client's parent might be facing when he or she makes the first call to your agency.

Exercise Three   Intake of an Infirm, Older Person

Instructions: Using the blank form in the Appendix entitled "New Referral or Inquiry," develop a client who is an older person, a person over 80 years of age. In this case, a child or close friend or neighbor would be calling on behalf of the client. A doctor may have referred the caller to you, or the caller may have felt the client needed help and sought your services without a referral. Typical problems confronting frail, older adults are problems with self-care, independent living problems, untreated medical conditions, malnutrition, and depression and anxiety.

In developing your client and your client's problem, read the *DSM-IV* and look at books and articles on specific problems common to older people. Look at the chapter in the companion textbook, *Fundamentals for Practice with High Risk Populations,* for information on how your client and the concerned caller might be feeling and what issues or problems the caller might be facing when he or she makes the first call to your agency.

Chapter 15
The First Interview*

INTRODUCTION

You have spoken to the person by phone, and you have arranged for the person to come into the office for an interview. This is the first interview for the client. Even if he or she was at one time a client of the agency, we will assume the case would have been closed for sometime.

The purpose of this interview is to establish basic information about the person, such as:

- Strengths, including external support systems, talents, successes, capabilities, and positive attitudes
- Weaknesses, including gaps in the external support system, lack of experience or information, negative attitudes, and events the client defines as failures
- Current problems that caused the client to seek help *now*
- Potential problems
- A sense of who the client is

YOUR ROLE

You have three tasks to accomplish in this first interview.

First, you must listen and convey an accurate understanding of clients' perceptions about themselves and their problems. When you convey this understanding, it does not mean that you necessarily agree with them, but it does mean that you have heard them accurately. In order to do this well, you need to allow clients to proceed in their own words. As they talk to you about what led to their seeking help, you can reflect back their feelings and perceptions about their situation, responding to feelings and to content. In this way, you sort out with the client what is important.

Second, you formulate a professional understanding of what it is the client is experiencing and what this person will need while being served by your agency.

---

* This chapter is based on *Where to Start and What to Ask* by Susan Lukas. Adapted with permission of W. W. Norton & Company, Inc. Copyright © 1993 by Susan Lukas.

Finally, you should strive to establish rapport with clients so that they feel comfortable with you and with your agency. Some people, no matter how hard you try, will never warm to the interviewer, but most people will respond positively to a worker who is warm, genuine, and empathetic.

THE CLIENT'S UNDERSTANDING

In most cases, clients have recognized the need for help; but in a few cases, clients may feel they do not need to be in your agency. Some clients are mandated by the court to seek help or face jail or the permanent removal of their children. In situations where clients feel forced to come to your agency, you may encounter hostility. In either case, you must indicate that you have heard all their concerns about being there. You can convey this through your ability to reflect back how the client is feeling about being in the agency.

Even when the clients believe they need help, they may not be clear about what their problems are or how the agency can help them. They may be clear that the current situation is painful, but unclear about how to describe it. They may know things seem out of control, but be unable to describe the impact their situation is having on them emotionally. They may hope that there is help available without understanding what kinds of resources there are or how these resources could help them specifically.

PREPARING FOR THE FIRST INTERVIEW

If you did not perform the telephone intake, you will want to look at the intake material that is available before the client arrives. As you do this, ask yourself what more you should know about the client's difficulties. What details need to be clarified? Where are there particular gaps that need to be filled in?

If the client was in your agency before, read past records to fill out the picture of this person. Look at past medical difficulties, medications the client might be on, or medications that were prescribed previously while the person was in treatment.

As you begin to form a picture in your mind of this person, remember that the information you have about this client was collected by others who saw the client under other circumstances. This may not be the whole picture, and it may not be an entirely accurate picture. Rely heavily, therefore, on your own insight and your own competence to form an accurate picture of the client.

At one time, there was a woman who was seen by various case managers, physicians, therapists, and psychologists who conducted testing. The woman had a problem with anger and had been asked to leave the home of several family members where she had been staying. At last she was residing in a group home. A psychologist was asked to do an evaluation for possible organicity (problems in the brain that would cause anger and loss of control).

In reading the chart with the many records in it, the psychologist came across an early note by the case manager: "Client created a scene at the local Giant food store last week over the fact that another customer took the last head of lettuce as she was reaching for it. Crisis intervention was called. Client was taken to her home." About a year later, in another note, a therapist noted, "Client made several scenes in the past at the Giant where she shops. Apparently she gets upset over the fact that there is not more produce in the store.

Management has called Crisis Intervention." Still later a new case manager wrote, "Client apparently creates violent scenes at the food stores when she feels there is not enough produce. Crisis was called on several occasions. Will advise client to stay away from food stores." Finally, several years later, a doctor prescribing medication noted, "Client was barred from shopping at any local food stores several years ago because of violent outbursts of rage over a lack of store items she meant to purchase. These outbursts resulted in contacts with Crisis Intervention and indicate a difficulty with anger control and poor communication skills." Between these individual notes in the chart, there was no other reference to outbursts at the food store. It appeared, from looking at the chart carefully, that each person who mentioned the incident was summarizing the previous note and magnifying it in the process. Think how differently you would approach this client if the first note you read was the doctor's note as opposed to the original case manager's note.

In this example, you do not know if the client changed her story each time she met someone new or if the workers read the charts and records and misinterpreted the information. There are other reasons why the notes you read may not be accurate. The person who wrote a note might have been hurried in her assessment. She might have felt hostility or prejudice toward the client for some reason. The note may have been made by someone who was inexperienced in interviewing. For all these reasons, you will have to rely on your own insight and competence in doing your assessment.

If you see inconsistencies in the previous history, make a note of them for further exploration.

YOUR OFFICE

Most case managers have an office or place where they see clients. Sometimes case managers share an interview room. Look at your office or interview room. Be sure it is a place you would feel comfortable in while confiding in another person. Is it warm and comfortable or utilitarian? Are there comfortable chairs? Is it free of harsh lighting? Are the walls attractive?

It is probably best not to have personal pictures sitting about because you cannot be sure how your clients will view these or what meaning they may find in them. A picture of a happy, smiling 3-year-old may be upsetting to a person whose children were just removed from the home or to someone who is struggling to find shelter for her own 3-year-old. It can create a barrier, and you are seeking to diminish barriers.

In your office, there needs to be a comfortable place for the client to sit facing you. You want her to be able to talk to you in a normal tone of voice, but you do not want her to feel crowded by your presence.

MEETING THE CLIENT

From the very beginning, you want clients to know that you respect them, that you wish to be helpful, and that you will be relating to them as a professional, and not as a social acquaintance. The interview begins with your first introduction.

1. Begin by going to the waiting room to meet your clients. Do not make them find their way through the halls to your office.
2. Introduce yourself as Ms. or Mr. _____. Or say, "Hello, my name is Jim Pellam." Do not say, "Hi. I'm Jim" or "Hello, my name is Jim." You did not spend all those years in school to earn this degree in order to be simply "Jim."

3. Make a mental note of your first impressions. How do the clients respond to your greeting? What do they say first? How do they look?

4. How do the clients react to your office? Do they seem comfortable? Do they appear to feel awkward? Do they readily sit down or wait to be invited to do so?

5. If clients start talking, show interest in what they are saying. Often the first things the client tells you will hold the most significance.

6. If clients ask about your credentials, tell them about these matter-of-factly. It is part of informed consent for clients to know who is seeing them There is no need for you to sound defensive. Do not go into personal details about yourself, however. If a client insists on knowing if you have ever had children or if you are old enough to know how to help her, point out respectfully that the purpose of her visit is to understand the issues and problems she is experiencing.

7. Describe the agency and explain its purpose to clients who are unclear about it. Some individuals come to a case management unit and expect to see a doctor or a psychologist. Give information about the types of professionals that staff your agency and what they do.

8. Make certain that you or someone else has described payment arrangements to the client.

9. Make sure that you or someone else has explained confidentiality and the limitations of confidentiality to the client. Clients need to know that their diagnosis may go to their insurance company. It is not necessary to go into every exception regarding confidentiality, but let them know that under circumstances where they might be in an emergency, information may be shared.

TAKING NOTES

It is all right to take notes during your interview. For one thing, you are collecting very basic information and you want to ensure that it is accurate. Let the client know that you are taking notes to make certain that you have accurate information.

During other contacts with the client, jot down significant phrases or information. You can reconstruct your contact in short notes after the client has gone.

COLLECTING THE INFORMATION

Allow your clients to tell their story in their own way, using their own words and expressions. Help them to begin talking about why they are here by asking an open question such as "Tell me a little bit about why you are here today" or "Can you tell me something about what brought you here?"

While the client is talking, remain emotionally neutral. Do not recoil in horror or gasp or squeal with delight. Do not tell the client how you would have felt under the circumstances. This is not about you. While the person answers your question, reflect back the content and feelings you hear.

If clients come from a different race, religion, or culture or have different values from yours, be aware of that without judging them. As we have seen, some case managers are tempted to judge clients using themselves as the standard. In other words, these case managers see themselves as the standard against which everything should be compared. By doing so, they miss the unique circumstances and characteristics that make the client a separate person. This diminishes the case manager's ability to be truly helpful.

ASKING FOR MORE CLARIFICATION

During the interview, ask for clarification. Use primarily open questions ("Can you tell me a little bit more about your father?" or "Could you describe your relationship with her before you were divorced?"). It is all right to ask closed questions if you need further clarification ("Bill is your boss?" or "You lived there how many years?"). Avoid too many closed questions so that your interview does not take on the tone of a grilling.

Avoid "why" questions as much as possible. Even when you have taken special care to ask them respectfully, a person may experience them as prying. Sometimes a "why" question actually asks clients to give an understanding of what motivated their actions or the actions of others. You may be asking for a level of insight clients have not yet developed. In that case, they can only feel incompetent and uncomfortable.

Sometimes clients know why they behaved in a certain way or why something happened, but the reason is upsetting to them or they are having trouble recognizing and talking about it. A "why" question can make them think that you will probe for answers and push them to talk about things they are not ready to discuss.

Finally, clients may tell you a great deal more than they had intended to tell the first time. They may go home upset with themselves and embarrassed by the amount of self-revelation in which they engaged. It may be so uncomfortable to them that they do not return to continue with your services. In that case, it is possible that you have intruded into the client's personal information too quickly. As the person conducting the interview and, therefore, the person with the most power, you have an obligation to protect the client from this kind of intrusion. A good way to protect clients is not to go too far beyond what they appear comfortable talking about.

Intrusions and discomfort of this sort can be avoided if you recognize from the very beginning that *the facts and circumstances of the client's life and problem belong exclusively to the client!* Clients are under no obligation to tell you more than they feel comfortable revealing. This means that you focus on the information clients can give you freely without probing and without discussing feelings and motivations.

WHAT INFORMATION TO COLLECT

The most important piece of information is to understand why the client is here now as opposed to last week or last month. Some of this information may be on the phone inquiry sheet, but your task in this first interview is to develop that information more completely. The reason the client has come to the agency now is often referred to as the "presenting problem."

In addition to the presenting problem, you want to understand the extent to which this problem has interfered with the client's ability to function socially, occupationally, and personally. Is this person able to work? Are the client's most important relationships feeling any strain? Is the person taking care of personal hygiene and other needs? You might ask questions like the following:

"Can you tell me something about how things are going at work?"
"Could you describe how things are at home?"
"Can you give me some idea of how this has affected your daily routine?"

Does the client have a support system or seem to be all alone? Individuals who are alone, or who perceive they are isolated, are at greater risk for stress and suicide than those who have a good support system in place. You might ask such questions as:

"Tell me a little bit about your family."

"Tell me something about your friends."

"Can you describe what you do in your spare time?"

CLIENT EXPECTATIONS

No service or treatment plan is entirely useful unless the client has participated in developing the plan. During this first interview, ask clients what it is they would like from your agency. You might ask them questions such as:

"Can you describe how you think we might be able to help you?"

"Tell me a little bit about the services you had in mind."

"Can you share with me some of your ideas about services you would like from us?"

Often clients do not know what services are available or what services they need. Together explore what your agency has to offer, and describe various alternatives for clients to consider. At the end of this first interview, you and the client need to have developed a tentative plan for services.

USING AN ASSESSMENT FORM

Many agencies have long forms that spell out what questions need to be answered. When you use such a form, it is possible to begin to sound like an interrogator. You may be asking, "Name? Address? Phone number?" and then move on to more involved questions such as "When were you married? How many children? Names and ages of your children?" Where is the warmth and empathy in such questioning?

The assessment form does contain many of the questions you need to ask to form a more complete picture of the person. If clients seem unwilling to answer any of the questions, move on to others in a matter-of-fact way. Do not try to persuade clients that they should answer something they feel uncomfortable discussing. Chances are this information will readily come out as you establish rapport.

The assessment form is simply an outline of what is important. If your intention is to fill in all the blanks on the form and close the interview, you have not really conducted an adequate interview. Most forms contain space for your comments or summaries. It is expected that you will talk with the client, that you and the client will discuss the situation, and that the client will volunteer additional information. Most forms soliciting information from clients are focused on the clients' problems and deficits. In order to round out a complete picture of the client, you need to look for strengths. Take the time to do so; and remember, this can happen only when the person feels comfortable with you. If you sound like a machine, the client will not connect with you at all.

WRAPPING UP

There are tasks that you want to complete toward the end of the interview.

1. Near the end of the interview, be sure to ask clients if they have any questions. Answer these questions thoughtfully, as this is part of giving clients information they need to give informed consent.

2. Work with clients to define their problem in language they can understand. This is very important because it gives you and the client a mutually clear definition of the presenting problem.
3. Talk to clients about what they expect as a result of coming to the agency. Ascertain what their goals are for themselves, the sort of service they are looking for, and the expected outcome. You might say, "Tell me something about where you would like to be a month (or four months, or whatever) from now."
4. Give clients some information about what will happen next. If the case is to be presented to a panel or treatment team, tell clients that, and tell them how long it will be before they will have information about a formulated plan for them. If they must go on a waiting list, tell them that, and give them information about where they can get services more quickly. Let them know what will happen after this first interview with you. Never let the client leave wondering what will happen next.
5. If clients are to return to you in a set amount of time to discuss the implementation of the plan that has been developed with them (or for some other reason), be sure to give them a card stating the time and the date of the next appointment. If someone must bring them to their next appointment (a parent, guardian, group home worker), be sure that person is aware of the time and date as well.

THE CLIENT LEAVES

Rise at the end of the interview to indicate that the session is complete. It is always a good idea to walk the client back out to the waiting room.

Be aware of something social service workers refer to as the "door-knob syndrome" wherein a client begins to tell you something of great significance just as she is leaving. It might be that the person saved this information for last deliberately because it is painful and she did not want to discuss it in depth. In any case, let her know the session will not continue and that she should bring the subject up with the therapist or the next time she sees you. Do this in a warm and interested manner. Do not appear to scold.

Do not allow clients to leave your office if you believe they are a danger to themselves or to others. If what they choose to bring up at the end indicates to you that this is a possibility, you will need to explore that further or see that there is someone who will.

After the client has left, do not go to the receptionist to place the client's name in the appointment book and discuss the client with the receptionist; and do not use the client's name where other clients in the waiting room can hear. Do not discuss your session in the hall or another case manager's office where other clients can overhear your comments.

THE NEXT STEP

Having completed the initial intake with the client, it is time to begin to put together a chart for the client. Charts or files are kept in a particular order. The order of the contents is precise, making it easier to find information. Sometimes information is color-coded. For example, financial information may be on yellow paper and assessments on blue. It is important to maintain your charts in the order specified by your agency, to return them promptly to the records section of your agency, or to file them properly. In addition, it is important to make sure that the charts are locked up safely when you leave for the night and that no one has access to them but those who should. Finally, never remove the chart from the agency. The possibility that the information could be lost, stolen, or read by others is too grave.

FIRST INTERVIEW EXERCISES

Exercise One   Assessment of a Middle-Aged Adult

Instructions: Using one of the blank assessment or evaluation forms found in the back of the companion textbook, *Fundamentals for Practice with High Risk Populations,* develop further one of the clients for whom you did a phone intake. Choose the assessment form that fits your client's problems. You would develop details about the client's life and gather information relevant to the reason the client called the agency.

In developing your client further, piece together the circumstances you believe might be reasonable for a person who has this particular problem. Assign the client a socioeconomic situation, amount of schooling, and the other particulars. Develop as well your client's problem by again consulting the *DSM-IV,* if relevant, and books and articles on this specific problem. Look at the chapter in the companion textbook, *Fundamentals for Practice with High Risk Populations,* related to the assessment form you chose to use for your client. Think about how your clients might be feeling as they come in for the first time and what issues or problems clients might be facing for which they are looking for help.

Exercise Two   Assessment of a Child

Instructions: Using the assessment form for children found in the back of the companion textbook, *Fundamentals for Practice with High Risk Populations,* develop further the child for whom you did a phone intake. You would develop details about the child's life and gather information relevant to the reason the parent or guardian called the agency.

In developing your client further, piece together the circumstances you believe might be reasonable for a child who has this particular problem. Assign the child's family a socioeconomic situation, amount of schooling, and the other particulars. Develop as well the child's problem by again consulting the *DSM-IV,* if relevant, and books and articles on this specific problem. Look at the chapter on children in the companion textbook, *Fundamentals for Practice with High Risk Populations,* and think about how you, the parent (or guardian), and the child might be feeling as the child comes in for the first time and what issues or problems the family might be facing for which they are looking for help.

To complete this assignment, you may use the chapter on children's mental health or the chapter on mental retardation found in the companion textbook, *Fundamentals for Practice with High Risk Populations.*

Exercise Three   Intake of an Infirm, Older Person

Instructions: Using the assessment form in the back of the companion textbook for senior citizens, further develop your client who is an older person, a person over 80 years of age. In this case, a child or close friend or neighbor may be present when you do the assessment and, depending on the condition of the older person, may give you most of the information. In addition, you may need to go to the person's home to meet with him or her because the client is too infirm to come to the office.

In developing your client further, piece together the circumstances you believe might be reasonable for an older person who has this particular problem. Assign a socioeconomic situation, amount of schooling, and the other particulars, such as marriages, number of children, former occupations, and interests. In developing your client's problem, read the *DSM-IV*, if relevant, and look at books and articles on the specific problem your older person appears to have. Look at the chapter in the companion textbook, *Fundamentals for Practice with High Risk Populations*, for information on how your client and the concerned caller might be feeling and what issues or problems the caller might be facing when he or she meets with you for the first time.

Exercise Four   Creating a File

Instructions: At this point you need to create a file on the people you are seeing as clients. Create a separate file folder for each client you intend to follow throughout the remainder of the course. Place the client's name on the tab of the file folder—last name and then first name, so that the cases can be filed alphabetically by their last names. Place the New Referral or Inquiry form on top, followed by the verification letter you sent. Under that, place the long assessment form you used, clipping the pages of that form together.

✗ Chapter 16
Using *DSM-IV**

INTRODUCTION

Is *DSM-IV* Only a Mental Health Tool?

Many students wonder why they need to learn about the *Diagnostic and Statistical Manual of Mental Disorders (DSM-IV)* when it appears to be a tool used exclusively by mental health practitioners. Actually the *DSM-IV* is a valuable tool you will use in many different settings. Although the majority to clients in the broad human service system will not have mental disorders, there are some places where *DSM-IV* helps to define what the client is experiencing and what that person needs. For instance, clients who come to agencies as victims of abuse or assault often suffer from posttraumatic stress disorder. Workers in agencies dealing with the problems of growing older will encounter people who have a dementia or symptoms that resulted from a stroke or other long-term, debilitating illness. Those who work with children in a variety of settings will encounter children who have learning difficulties. A familiarity with the language and process of the *DSM-IV* allows you to participate in planning for the client more competently.

Today, with deinstitutionalization of the mentally ill, those with mental disorders come for services at many social service agencies; and we see more people who have more than one problem. You might be working at a shelter for victims of domestic violence and do an intake for a woman who is also a person suffering from bipolar I disorder. Clients no longer fit into neat boxes with no overlapping problems. For that reason, it is important to be familiar with this system.

The *DSM-IV* is the language of insurance companies and other funding sources with regard to any of the behavioral treatments such as drug and alcohol treatment or mental health and mental retardation. In addition, the *DSM-IV* contains information about situations and problems that may not constitute a mental disorder but may be the focus of attention in a clinical setting. Many of these situations come to the attention of social service agencies not equipped to treat them. You will need good information to make sound referrals.

It is for these reasons that your ability to understand the *DSM-IV* and your acquaintance with the various classifications of mental disorders will enable you to be more conversant with others in the field and to know more definitively a mental disorder when you encounter one.

---

*Note that *DSM-IV* was revised in 2001 to become *DSM-IV-TR*.

Cautions

Having spelled out why the *DSM-IV* is important in human service practice, it is equally important to understand that most people who come for services in social service agencies are not suffering from a mental disorder. The *DSM-IV* cannot be used to help you understand every client. If you try to give a psychiatric label to everyone you see, you will unnecessarily burden individuals who are well, but grappling with life events and disruptions.

In addition, the *DSM-IV* comes from the medical model. That is, the model suggests that individuals are labeled with an illness and are then treated as sick. This is a view of the client that can cause you to lose sight of the fact that the client has strengths and successes. Although a diagnosis is useful to clinicians in providing treatment, to case managers it can have the subtle effect of diminishing the client as a whole person.

The agency where you work will have policies and guidelines about using the *DSM-IV;* and many agencies will not rely on the manual at all. When you must rely on this manual, be very careful about categorizing people and allowing the person's diagnosis to color your complete understanding of the person.

Who Makes the Diagnosis?

You are not studying the *DSM-IV* to make diagnoses. A physician does that, or sometimes a senior staff person with a Ph.D. Nevertheless, the *DSM-IV* contains a language that is universally understood. Your experience with this language and with the mental disorders in the manual will facilitate your communication and reports to those responsible for giving the diagnoses. It could happen, on a rare occasion, that a harried emergency room physician with a waiting room filled with medical emergencies will turn to the emergency worker from a social service agency and ask that the worker give a provisional diagnosis to facilitate admission to the hospital (where the diagnosis will be reevaluated in less pressing circumstances).

It is important for you to keep in mind that additional clinical information is *always* needed to help round out the picture and make the best diagnosis and treatment plan. Much of that additional information in many settings will come from your histories and notes.

This chapter is based almost entirely on the work of Anthony L. LaBruzza (1994), whose book *Using DSM-IV: A Clinician's Guide to Psychiatric Diagnosis,* gives excellent background to how we arrived at the DSM-IV and how to use it effectively.

BACKGROUND INFORMATION

Up until the 1600s, physicians used a patient's horoscope to diagnose mental disorders. Medieval physicians looked at the four humors to account for differences in human personality and temperament. The humor that predominated accounted for the patient's disposition—with blood accounting for a happy temperament; choler contributing to a fiery, competitive temperament; phlegm resulting in a cold, delicate disposition; and bile causing melancholy.

Psychiatry Attempts to Classify Mental Disorders

In colonial times, most mentally ill individuals were managed at home by their families. Many were abused and exploited or were confined to workhouses and almshouses where varying theories about the reasons for their illnesses caused harsh treatment in most cases. Between 1800 and 1860, there were a number of people concerned with placing those with mental illness in "asylums" where a more humane approach and more respect for the patient would be the rule. Such treatment was referred to as "moral treatment" (LaBruzza, 1994). Dorothea Dix was active in this movement; and when her attempts to start a federal asylum program failed, she became immensely instrumental in the founding of state hospitals in Pennsylvania and New Jersey, which bear her mark to this day.

Mental illness was still little understood; and in the census of 1840, people were classified as either sane or "idiocy/insanity." The shift from the asylum to treatment, research, and education occurred in the late 1800s and early 1900s. At that time, research was beginning to provide a clearer picture of the anatomy of the brain, and the diagnostic system became more refined so that by 1880 there were seven categories of mental disorder.

Diagnosis continued to be the focus of research. Wilhelm Greisinger (1817–1868) in Germany looked at the mental disorders as diseases of the brain, an organic view. Another German, Emil Kraepelin (1855–1926) looked at syndromes or collections of symptoms and made statistical records of the symptoms the patients exhibited, the course of their diseases, and the outcomes. His goal was to be able to accurately predict the outcome of a disorder for a patient based on certain combinations of symptoms. He used a behavioral and descriptive approach that made it easier for others to use his concepts.

There were others who imposed their views of the brain and nervous system in creating a diagnostic classification system. Most influential in the United States was a Swiss-born psychiatrist, Adolph Meyer (1866–1950). Mental disorder in his view was a response to psychosocial stressors. This view was widely accepted as individuals drafted into the military during both world wars appeared to break down under the stress of combat. Had his view continued to be influential, mental illness would have been seen today as an adaptive response. Instead, mental disorders gradually came to be seen as discrete psychiatric diseases.

In the 1920s, the American Psychiatric Association (APA) decided to find a way to standardize the medical terminology psychiatrists used. A national conference in 1928 looked at how diseases were named. The classification system that emerged focused only on the most severe forms of mental disorders, those that would most likely cause the patient to be institutionalized. The classification became broader when World War II veterans returned with less severe disorders. In the 1940s, there were ten types of psychoses, nine neuroses, and a remaining seven disorders related to behavior, intelligence, and character. In 1952, the APA published the first *Diagnostic and Statistical Manual*.

In 1965, the APA, in an attempt to keep up with international changes in the way mental disorders were classified, revised the original manual and brought out the second edition, *DSM-II*. It was in this manual that there seemed to be a return to the Kraepelian descriptive model for diagnosis. Those who did the revisions eliminated terms that implied a particular theory of etiology (or cause) for the disorder. This successfully did away with Meyer's idea of seeing mental disorders as a response to stress. Nevertheless, psychoanalytic terminology remained because psychoanalysis was still quite popular and influential among psychiatrists.

The 1950s, 1960s, and 1970s

At this point, the manual was still quite unreliable. Psychiatrists would give different diagnoses to the same symptoms, making research impossible. Anthony LaBruzza (1994) stated, "[T]he possibility that two psychiatrists would agree on the same diagnosis in the 1950s and 1960s was nearly random." In the 1960s, psychiatry was out of favor with the public as famous court cases pitted psychiatrists against each other in what appeared to be a nebulous theoretical system and movies like *One Flew Over the Cuckoo's Nest* introduced moviegoers to the possibility that institutions were punitive and the staff in such places were not much healthier than the patients. This was a time when all authority was challenged, and a number of books challenged the authority of psychiatry, particularly Thomas Szasz's book *The Myth of Mental Illness*. There were many who saw psychiatry and psychiatric diagnoses as stigmatizing and wielding undue social control.

In addition, insurance companies began to cut back on the amount of psychiatric care for which they were willing to pay, in part because the diagnosis of mental illness was unreliable and there seemed to be no consensus on the best treatments. No studies had been conducted to determine which illness responded to which treatment.

Psychiatry Becomes More Medical

The third edition of the manual, *DSM-III*, came out in 1980. Every edition of the manual since *DSM-III* has been an expansion or refinement of that document. This manual relied on a more medical research–oriented model of disease, and it also relied more heavily on the Kraepelian use of descriptions. In addition, it was no longer slanted toward psychoanalytic descriptions or causes; in fact, causes were, for the most part, left to research to determine. Responding to the concerns voiced about psychiatry, the third edition of the manual contained fourteen discrete and specific mental disorders with very explicit descriptions. These descriptions had operational criteria that allowed them to be measured statistically. All references to unconscious motives were removed, and the clinician based the diagnosis strictly on what could be seen.

The changes in the third edition of the manual could be summarized as follows:

1. There was every attempt to use clear English, and not mental health scientific jargon.
2. Disorders were labeled, and not people.
3. *Patient* was dropped in favor of words like *person* or *individual*.
4. The manual was tested for reliability for the first time by clinicians using it in the field.
5. A multiaxial system was adopted to give a fuller diagnostic picture of the person.
6. Decision trees were included to help the physician rule out similar disorders and narrow the diagnostic choice to one.
7. The words *disease* or *illness* were dropped in favor of the word *disorder*.
8. All the pet theories about causes of disorders were eliminated.
9. Each disorder had a working definition that contained operational criteria (criteria that could be measured).

After the publication of *DSM-III*, psychiatrists were far more likely to make the same diagnosis for the same set of symptoms. This enabled research to be done more effectively, particularly field trials of medications that treated specific psychiatric symptoms. In other words, it became more likely that practitioners would all agree on the diagnosis for certain clusters of symptoms, regardless of where they were practicing. If everyone was seeing the same thing when they looked at a cluster of symptoms, then it was possible to treat that cluster of symptoms in various ways to determine the best approach to alleviating the symptoms. Now clinicians could communicate reliably in a common language about diagnoses. This common language facilitated good research. Pharmacy companies supported this research for products they developed for these specific disorders.

How We Got to *DSM-IV*

With all the field testing that took place as a result of *DSM-III*, revisions were inevitable. Thus, in 1987, *DSM-III-R* (or revised) came out; this edition included twenty-seven new categories and revisions to some older diagnoses. The number of categories went from 265 to 292. An appendix contained further categories requiring additional research.

The *DSM-III-R* made another important shift that will affect your work with the document. It moved from the *monothetic diagnosis* to the *polythetic diagnosis*. The old *DSM*s used the monothetic diagnosis. They gave a series of symptoms that constituted a disorder, and unless all of them were present you could not use the diagnosis. This meant that a diagnosis was only as useful as the least useful item in the series of symptoms. In a polythetic approach, the series of symptoms is given, and the patient must have several of them but not all of them. This improved the reliability of diagnoses.

Another important shift was the move to give a patient more than one diagnosis if the patient met the criteria for more than one. Before, the clinician had to choose the diagnosis that was most obvious or urgent. Other diagnoses that coexisted with the first diagnosis, or were, perhaps, part of a larger clinical problem, were not mentioned. This narrowed the clinical picture of the patient. Now a fuller clinical picture was possible. The *DSM-III-R* also lined up with the new version of *International Classification of Diseases (ICD-10)*, which made it easier for U.S. clinicians to talk to clinicians internationally.

The *DSM-IV* attempted to make as few changes as possible and made certain that the changes that were made were based on good research with empirical results. In order to get the empirical basis for changes, the work committees (those committees working on various classifications) systematically reviewed the literature for different diagnostic categories, reanalyzed previous data, and conducted field trials to make certain the diagnoses were reliable in many different settings and in many different types of clinical work. The *DSM-IV* also did away with all sexist language.

USING THE *DSM-IV*

What You Will Find in *DSM-IV*

The following are some of the features you will find in the fourth edition of the manual.

1. Every disorder has a name, numerical code, the criteria needed in order to give the diagnosis, the subtypes of the disorder, the specifiers, recording procedures, and examples that illustrate the disorder.
2. Associated features and associated disorders may include such items as clinical features that may be present but are not always seen in the disorder; disorders that precede, often co-occur, or generally follow the disorder in question; typical laboratory findings; physical signs and symptoms; and typical medical conditions.
3. The typical age at onset and any cultural and gender-related information are provided.
4. The manual lists the prevalence of the disorder, the incidence, and the risk.
5. A description of the typical clinical course of the disorder is included.
6. Any complications that might be applicable to the disorder are given.
7. Typical predisposing factors discovered through research are listed.
8. The manual supplies family patterns if there are genetic or suspected genetic components to the disease.

9. Differential diagnoses or disorders that share similar symptoms are provided, along with information on how to make the distinction. (LaBruzza, 1994, pp. 57–58)

Making the Diagnosis

The patient receives a diagnosis along five separate dimensions, referred to as axes. Each axis gives different information about the patient, providing a more accurate clinical picture than would be possible with a single axis. This is called a multiaxial diagnosis. The diagnoses entered on each axis have both names and numbers, the numbers being useful for insurance and billing purposes. Each axis serves a different purpose. Figure 16.1 contains an outline showing what information is coded on each axis.

A Closer Look at Multiaxial Assessment

Axis I

It is assumed that the primary diagnosis or reason a person is seeking treatment will generally be the diagnosis appearing on Axis I. If there is more than one Axis I disorder, the primary diagnosis is listed first and is often qualified with the phrase "Reason for visit" or "Principle diagnosis" (unless the principle diagnosis is on Axis II) (LaBruzza, 1994). If the clinician does not use one of these phrases, the Axis I diagnosis is always considered to be the primary diagnosis. The following are the disorders coded on Axis I:

1. Disorders usually first diagnosed in infancy, childhood, or adolescence. Mental retardation is the exception and is coded on Axis II

Axis 1	All clinical syndromes listed in DSM-IV are coded on this axis *except* personality disorders and mental retardation.
	Includes developmental disorders
	Other conditions that might be a focus of clinical attention
	V71.09 No diagnosis on Axis I
	799.9 Diagnosis deferred on Axis 1 (meaning too little time or information to establish a diagnosis)
Axis II	Personality disorders
	Mental retardation
	Significant maladaptive personality traits
	Habitual defense mechanisms
	V71.09 No diagnosis on Axis II
	799.9 Diagnosis deferred on Axis II
Axis III	All general medical conditions that are relevant to planning and understanding the patient's diagnosis
	International Classification of Diseases (ICD-10) codes can be used here.
	None (meaning no medical conditions)
	Deferred
Axis IV	Psychosocial and environmental problems that affect the prognosis, management, or treatment of the case
Axis V	The rating on the Global Assessment of Functioning (GAF) scale, which is usually a single number between 1 and 100 indicating the current level of functioning the patient possesses

Figure 16.1 Dimensions used in multiaxial diagnosis

2. Delirium, dementia, amnestic and other cognitive disorders
3. Mental disorders due to a general medical condition not elsewhere classified
4. Substance-related disorders
5. Schizophrenia and other psychotic disorders
6. Mood disorders
7. Anxiety disorders
8. Somatoform disorders
9. Factious disorders
10. Dissociative disorders
11. Sexual and gender identity disorders
12. Eating disorders
13. Sleep disorders
14. Impulse control disorders not elsewhere classified
15. Adjustment disorders
16. Other conditions that may be a focus of clinical attention

When there is no diagnosis on Axis I, the clinician writes V71.09 on the axis.

Axis II

Axis II is used to code personality disorders and mental retardation. This was done to be sure relevant personality factors would be part of the entire diagnosis. It also makes mental retardation a separate factor on a separate axis in the patient's general diagnosis. *Of all the disorders listed in the DSM-IV, only mental retardation and the learning disorders require diagnostic testing before the diagnosis can be given.* If a patient meets the criteria for more than one personality disorder, all of them should be coded on Axis II. In addition, the clinician can write in any significant maladaptive personality traits or habitual defense mechanisms on this axis. These items have no code number.

The absence of an Axis II disorder is coded V71.09. If the practitioner suspects there may be an Axis II disorder but is not sure, the deferred code, 799.9, is used. You will find that insurance companies are not pleased with Axis II diagnoses due to the amount of costly treatment required. This treatment can often be intense and time-consuming.

There are eleven categories of personality disorder. They are:

1. Paranoid
2. Schizoid
3. Schizotypal
4. Antisocial
5. Borderline
6. Histrionic
7. Narcissistic
8. Avoidant
9. Dependent
10. Obsessive-compulsive
11. Personality disorder not otherwise specified

In addition, borderline intellectual functioning is also coded on Axis II (V62.89). When there is no diagnosis on Axis II, the clinician writes V71.09 on the axis.

Axis III

Often, individuals with severe mental disorders will also have general medical conditions that affect the prognosis, treatment, and even the understanding of the patients' situation.

Any general medical condition that is relevant should be coded on Axis III, along with any medical history that may be relevant to the current problem.

Axis III is not meant to indicate the mind and body are entirely separate entities. Axis III is separate in order to be sure the full picture of mental and medical disorders is recorded. People with medical conditions often feel less well psychologically. Axis III also alerts the physician to the possibility that the patient is on medications that might interfere or interact negatively with psychotropic medications that might be prescribed for the mental disorder.

In some cases, it is the medical condition that has caused the mental disorder. If this is the case, the medical condition is recorded on Axis III, and the mental condition is recorded on Axis I with the phrase "due to . . ." (for example, Axis I might read "major depressive disorder, single episode, due to hypothyroidism," while Axis III would read "hypothyroidism").

If no medical condition exists, write "Axis III: None." If it is suspected that there may be a medical condition but there is not enough information, it is coded "Axis III: Deferred." The clinician can note in writing any significant symptoms or physical signs that were observed that need further evaluation.

Axis IV

Axis IV should list any psychosocial stressors or environmental problems that appear to have an impact on the conditions noted on Axes I, II, and III. The clinician looks at what has happened to the client, particularly in the last year. Some stressors happened a number of years ago. For instance, a diagnosis of posttraumatic stress disorder (309.81) on Axis I may have Vietnam War on Axis IV as a stressor that is affecting the current condition of a Vietnam war veteran.

In an attempt to categorize most pschosocial stressors, a list of V codes was developed outlining broad general catoegories of psychosocial stress a person might experience. These are:

1. Problems with primary support group

   Childhood (V61.9)
   Adult (V61.9)
   Parent-child (V61.20)

2. Problems related to the social environment (V62.4)
3. Educational problems (V62.3)
4. Occupational problems (V62.2)
5. Housing problems
6. Economic problems
7. Problems with access to health care services
8. Problems related to interaction with the legal system/crime
9. Other psychosocial and environmental problems

If stress from a specific stressor is the main reason a person is seeking help, the V code for the stressor should go on Axis I.

Much psychosocial stress has to do with problems in the person's support system. This can include births, deaths, separation, divorce, remarriage, abuse, neglect, and significant illness. When looking at the person's social environment, you might find social isolation, lost friendships, retirement, relocations, and cultural differences as contributors to the current difficulty.

The "other" category can be a place to record natural disasters and catastrophes (for instance, floods or earthquakes) or problems related to professional caregivers (for example, a nursing assistant who is rough and seemingly threatening). The term *other* is used for any stressor that does not fit into the other categories. After the category is listed, the clinician should write out the specific stressors within each category (unemployed, best friend was killed, mugged, and so forth).

There are positive stressors such as weddings or the birth of a child. These are positive stressful events and are not coded on Axis IV unless they create a clinical problem for the patient.

Axis V

This axis is reserved for the patient's number on the Global Assessment of Functioning (GAF) scale. How well a person functions or does not function will affect any treatment plan developed and may affect insurance payment, research options, and the overall assessment of the person. A single rating of how the person is functioning at the time of the evaluation is given on Axis V. You will find the GAF in your *DSM-IV* book. The number assigned the patient should reflect the person's "current level of psychological, social, and occupational functioning at the time of the evaluation," according to Anthony LaBruzza (1994, p. 79). He further points out that "[i]mpairment due to physical, and environmental limitations is excluded from Axis V" (LaBruzza, 1994, p. 79).

The number given on Axis V is a number from 1 to 100. Zero indicates that there has not been enough time to adequately assess the functioning. The scale is divided into 10-point segments, with 1 to 10 being the lowest functioning and 90 to 100 being the highest. Most individuals requiring inpatient hospitalization usually have a GAF score of 50 or less. The scale has been criticized for making it possible to a greater degree to abuse reimbursement, forensic, and disability situations.

MAKING THE CODE

All the disorders in the *DSM-IV* have a numerical code. The code has three whole numbers followed by a decimal point and one or two additional numbers. The form of the codes looks like this: XXX.XX. The last two numbers in the code give the diagnosis more specificity.

Let us take an example. A person comes in with an obvious depression. In *DSM-IV* terms, this is called a "major depressive episode" (296._ _). The number 296.21 indicates the person's condition is mild, 296.22 means moderate, 296.23 means severe but without psychotic features, and 296.24 is severe with psychotic features. A person in partial remission would get 296.25, while a person in full remission would get 296.26. We know that the Axis I diagnosis will be 296._ _. By choosing the proper digits to follow the decimal point, we create a more accurate clinical picture of this person.

The fifth digit is generally used to identify four things:

1. Subtypes
2. Modifiers
3. Course of the disorder
4. Severity

Subtypes

When the clinician codes a diagnosis, the manual may require that the clinician "specify type." This refers to subtypes within a specific diagnosis. For instance, the diagnosis may be a delusional disorder (297.1), but there are seven distinct types of delusional disorders, including jealous, persecutory, grandiose, and so forth. The fifth digit is used to assign the subtype the clinician is diagnosing.

Modifiers

The manual may require that the practitioner "specify if" when coding a diagnosis. These are modifiers that allow the clinician to indicate if certain factors are present in this particular diagnosis. For instance, if the diagnosis is pedophilia (302.2), the practitioner will be asked to specify if the person is sexually attracted to males, females, or both (LaBruzza, 1994).

Modifiers for Past and Present

It is always understood that the diagnosis is a present condition. There will be times, however, that a past history of a particular diagnosis will be useful to note. A diagnosis from the past is modified with the phrase "prior history." In this case, you might write: "Posttraumatic stress disorder (309.81), prior history."

Modifiers for Course and Severity

The fifth digit is also used to specify the status of the remission or the degree of severity. Severity is usually either mild, moderate, severe, or psychotic. Each of these degrees of severity, specified by the fifth digit, indicate the intensity of the signs and symptoms and the degree to which the person's functioning is impaired. Mild means the person's symptoms just minimally meet the criteria for the disorder. Severe means the person meets the full criteria with intensity and is severely impaired in functioning.

Remission is usually either "partial remission" or "full remission." Partial remission means the person meets some of the criteria for the disorder, but is not any longer showing the full criteria that were present when the original diagnosis was made. You might also use this specifier to indicate a person who had been in full remission and is now showing some of the criteria again. Full remission means all the signs and symptoms of the disorder have disappeared, but the diagnosis is clinically significant at present.

Reason-for-Visit Modifiers

The *DSM-IV* made the shift toward multiple diagnoses. If the person fits the criteria for more than one disorder, then all of the disorders are listed. This means that the practitioner needs to identify the disorder that brought the client in for treatment or the disorder that will most likely be the focus of treatment. This is done by placing the phrase "principal diagnosis" or "reason for visit" beside the primary diagnosis.

List the primary diagnosis first with the modifier ("reason for visit" or "principal diagnosis"), then list the other diagnoses under that in order of importance. Generally, the primary diagnosis is on Axis I. On occasion, it will be on Axis II, in which case be sure to use the modifier on the Axis II diagnosis.

Provisional Diagnoses

When it is not clear what the diagnosis should be, the clinician needs to indicate that uncertainty. It may be that the client is uncooperative or impaired to the point that he or she cannot give much information. If the clinician has a strong idea of what the diagnosis probably is, a diagnosis is written with the term "provisional" after it.

There are some disorders that *must* initially be modified with "provisional." These would be disorders where there is a time lapse needed between episodes in order to be sure. For example, panic disorder (300.01) requires at least one panic attack, followed by at least a month of persistent worry about having more attacks. If the clinician sees someone who has what he or she believes is panic disorder and the panic episode occurred only two days ago, the client cannot be given the diagnosis with a degree of certainty. Thus the word *provisional* is used until enough time has passed to confirm or revise the diagnosis.

Not Otherwise Specified

There will be times when clients come in with most of the symptoms of a particular disorder and the practitioner feels he or she knows what the diagnosis should be, but the signs and symptoms do not quite fit the criteria as they are outlined in the manual. All the major classes of mental disorders have a "not otherwise specified," or NOS, category. The practitioner would write NOS after the diagnosis in this sort of case.

For example, a person might have symptoms that do not meet the full criteria for any one disorder, but meet some criteria for several different disorders, giving a mixed version of the disorder. Or you might encounter a person who has an obvious mental disorder, but it is not found in the *DSM-IV.* Perhaps it is a characteristic of the person's culture, or it is something being researched that has not been given a code yet. In another situation, the clinician may not know the cause of the disorder. The mental disorder he or she is seeing might be due to the medication the person is on, the general medical condition the person has, or the stress of the individual's life. This may need to be sorted out before giving a certain diagnosis. Finally, if the clinician lacks enough information to be sure of the diagnosis, he or she could use the NOS category. After getting good information, the diagnosis would be changed in this case.

V Codes

The *DSM-IV* lists "other conditions" that may be the focus of clinical attention. These codes begin with the letter V and are listed on Axis I. They refer to problems people encounter in life that might cause them to seek professional help. It might be bereavement (V62.82), school problems (V62.3), or any number of situational and relationship problems that have come to clinical attention. Sometimes after further treatment, a discrete mental disorder emerges and the diagnosis is changed at that point.

These V codes should not be confused with V71.09, which is the code used on both Axis I and Axis II to indicate there is no clinical condition present on that axis.

Unspecified

There is an "unspecified mental disorder" category that cannot be used when psychotic symptoms are present. The number is 300.9. This is used when there is some certainty that a mental disorder exists, but there is inadequate information to make a clear diagnosis. Later it can be changed to a specific disorder, unless it is found that the symptoms do not meet the criteria for any specific disorder.

Deferred

If there is inadequate information to make any diagnosis, the number on both Axis I and Axis II is 799.9, "diagnosis deferred."

CONCLUSION

The *DSM-IV* is a complex manual. It takes practice and good clinical skills to use the manual effectively. Nevertheless, entry-level individuals are being asked to understand the categories of disorders and discuss diagnoses with clinicians. In this course, we will begin to look at how you would use the manual in your work as a case manager. As you practice, you will find that you are clearer about how disorders are defined and how treatments are assigned.

EXERCISES ON USING *DSM-IV*

Instructions: See how many of the following exercises you can complete. These are designed simply to familiarize you with where things are in the *DSM-IV* and where conditions are coded on the five axes. When it comes to actual diagnoses, there will always be debate about what diagnosis to use. After discussion, in which you will no doubt cover many of the issues raised in real work situations, assign the diagnosis you feel is most appropriate.

1. It seems to you that Jim is having trouble in school. The teacher reports that he can read and speak well, but has trouble writing out his thoughts in a coherent and organized manner. You suspect the diagnosis is _____, and you know that _____ must be done to confirm it.

2. The doctor asks you to estimate a GAF for a woman who is being seen by you in the ER. She has been confused recently, according to her family; and today she left some pots on the stove and forgot to attend to them. The kitchen caught fire. In talking to her, she seems somewhat unconcerned or unaware of the gravity of the situation. She also seems unable to find the ladies' room when you direct her there, and you end up going with her and leading her back to the interview room. The abbreviation GAF stands for _____. You assign her a GAF of _____. You place the GAF on Axis _____.

3. A man comes in and indicates he is suffering from severe depression. He appears to have a flat affect and some tearfulness. In the course of the interview, you learn that he is an intermittent cocaine user. How do you code these two disorders? Think carefully!

4. A woman has breast cancer and has been depressed since the diagnosis was first given. Her family reports that she seems to be getting worse. How do you code these two disorders? Think carefully!

5. A patient is being admitted to the psychiatric unit, and the doctor is unclear whether she is seeing a personality disorder or a clinical syndrome. She needs to get the patient admitted quickly and knows the doctor on the floor will have the time to sort this out.

   What does the doctor most likely write on Axis I? _____

What does she most likely write on Axis II? _____

6. The psychologist laments to you that she knows the client has a dissociative disorder, but goes on to say that although it looks similar to dissociative identity disorder, she cannot quite see two or more distinct personality states.

   You think of what diagnosis?

7. For each of the following, indicate on which axis it should appear:

   Migraine headaches _____
   Housing problems _____
   Cancer _____
   Schizophrenia, paranoid type _____
   Borderline personality disorder _____
   A broken leg _____
   A divorce _____
   A lost job _____
   Severely impaired social and occupational functioning _____
   Dementia of the Alzheimer's type _____
   Domestic violence _____
   Mental retardation _____
   Inhalant intoxication _____
   Histrionic personality disorder _____

8. If you know that the diagnosis for schizophreniform disorder is an episode of the disorder lasting at least one month but less than six months, and you must make the diagnosis without waiting for the recovery, you would mark the diagnosis _____.

9. Describe two clients.

   Client #1 has 296.21, major depressive disorder, with melancholic features. Describe the symptoms, and tell the Axis on which this diagnosis will go.

   Client #2 has 296.33, major depressive disorder, with catatonic features. Describe the symptoms, and tell the Axis on which this diagnosis will go.

10. All personality disorders have as the first three digits _____.
11. All anxiety disorders have as the first three digits _____.
12. Depressive disorders have as the first three digits _____.

13. Schizophrenia has several subtypes. Describe a person whose diagnosis is 295.10.

14. How does this person differ from a person with a diagnosis of 295.30?

15. Assign a GAF to this person. A man is seeking counseling for depression following the death of his wife. He has been preoccupied and forgetful at work and finds himself crying alone at night. GAF is _____.

16. Assign a GAF to this person. A man tells you he has been anxious lately. He makes vague references to neighbors and a plan the neighbors have that worries him, but does not elaborate. He also mentions the need to use only stores that stay open 24 hours a day in order not to be involved in the neighborhood plan. He believes his neighbors have had an influence in Washington beyond what their numbers would indicate, but attributes it to their "plan." He is working and reports he recently received a raise. He has few friends and enjoys his job as his main source of socializing. GAF is _____.

17. Assign a GAF to this person. A woman was brought from her apartment by ambulance to the hospital emergency room after neighbors who were concerned about her called for help. She is unwashed, smells bad, and is mute. Attempts to communicate with her are in vain. She looks past the worker and does not appear to hear anything addressed to her. She is very thin and has bad breath. GAF is _____.

18. Assign a GAF to this person. A man is about to take his state psychology boards. He has made nearly straight As in graduate school. He is happily married and the father of two daughters. He plays tennis on the weekends and is an expert cook. He and his wife hold season tickets to the symphony, and he is a deacon in his church. He is complaining of rapid heart beat and sweaty palms. GAF is _____.

 **Chapter 17**

The Mental Status Exam

INTRODUCTION

The mental status exam (MSE) is based on your *observations* of the client. It is not related to the facts of his or her situation, but to the way the person acts, how the person talks, and how the person looks while in your presence. Mental status examinations can be abbreviated assessments done because someone appears to be in obvious need of hospitalization, or these examinations can be an elongated process that takes place over several interviews. The MSE always has the same content, and you write your observations in roughly the same order each time.

Although a formal MSE is usually done by a physician or psychologist, you will do an informal MSE in which you systematically look at the client's thinking process, feeling state, and behavior. You will want to understand the way the client functions emotionally and cognitively.

Much of the examination is done by observing how clients present themselves at the interview and the manner in which they spontaneously give information about themselves and their situation. The examination is not done separately, but is an integral part of the assessment interview. Questions that relate to mental status are framed as part of the overall assessment, and not as a separate pursuit. There will be times when you or a clinician might ask for psychological testing to confirm your evaluation of the client, but during your own MSE of the client, this is not done.

Some of the terms you will learn in this chapter are not necessarily words you will use in describing your clients and their appearance or behavior. This chapter is meant to make you familiar with the way some professional practitioners describe their clients. If you know these terms, you will be able to follow the notes or the discussion better.

WHAT TO OBSERVE

Your mental status examination of the client involves observations of the following:

General appearance   Cognitive functioning
Behavior   Intelligence

Thought process and content   Reality testing
Affect   Suicidal or homicidal ideation
Impulse control   Judgment
Insight

A good case manager is a good observer. You pick up many details about the client, all of which are relevant to understanding the client's mental status. In a sense, you watch for the most obvious and the most subtle visual and verbal clues as to who your client is. Use what you see and hear to give you direction in regard to what questions to ask.

HOW TO OBSERVE

Throughout the interview you will be noting how the client communicates verbally and nonverbally and how the client behaves. In addition, you will look at the content of the communication. You are looking at both *what* the client tells you and *how* the client tells it.

This means that as clients talk about why they came to your agency for services and what the main problems are that they are confronting, you are making some judgments about how they functioned in the past and how well they are functioning currently. You will note how the client tells his or her story. Is the person cooperative and friendly? Does he appear relieved and eager to talk to you, or is he mute, guarded, and uncooperative? Is she weepy and hesitant as she speaks, or is she forthright and stern? Does the person twist a tissue in her hands or rock back and forth in her chair, or does she use appropriate gestures? Does he relax during the interview or remain guarded and uncooperative?

At times you may need to assess clients' mental status through the observations of others who are close to them. Your clients may not always be able to tell you much about past events or functioning, and you will need to turn to others for that information. If there is no reliable source, you may not be able to perform a complete mental status examination that has a clear degree of certainty.

DOCUMENTATION

It has been suggested that you be able to back up your observations with both descriptions of the client's behavior during the interview and direct quotes made by the client in the interview. In this way, you carefully document your observations and your resulting conclusions.

DESCRIBING THE CLIENT

When you describe another person, be sure that your values and prejudices do not appear in your notes. Use adjectives that describe the client, but are objective. All editorial comments and value judgments should be omitted.

MENTAL STATUS OUTLINE

Anthony LaBruzza (1994), in his book *Using DSM-IV*, provides a good outline for the mental status interview. He points out that his outline is not meant to be followed precisely, but it does give the major points and a framework to determine what is important. The outline is shown in Figure 17.1.

The rest of this chapter provides terms to use and items on which to focus for each category on the examination. General terms are also included for your use.

The Mental Status Examination

1. General Description
    A. Appearance
        1. Dress and grooming
        2. Physical characteristics
        3. Posture and gait
    B. Attitude and interpersonal style
    C. Behavior and psychomotor activity
    D. Speech and language
        1. Rate
        2. Clarity, pitch, volume, tone, quality, and resonance
        3. Abnormalities
II. Emotions
    A. Mood
    B. Affect
    C. Neurovegetative signs of depression
III. Cognitive Functioning
    A. Orientation and level of consciousness
    B. Attention and concentration
    C. Memory
        1. Immediate registration, retention, and recall (a minute or less)
        2. Recent memory (a minute to days or weeks)
        3. Remote memory (weeks to years)
            a. Memory for recent past
            b. Memory for distant past
        4. Client's subjective report of memory difficulties
    D. Ability to abstract and generalize
    E. Information and intelligence
        1. Fund of knowledge
        2. Estimate of intelligence
IV. Thought and Perception
    A. Disordered perceptions
        1. Illusion
        2. Hallucinations
        3. Depersonalization and derealization
    B. Thought content
        1. Distortions
        2. Delusions
        3. Ideas of reference
        4. Magical thinking
    C. Thought process
        1. Flow of ideas
        2. Quality of associations

D. Preoccupations
      1. Somatic
      2. Obsessions and compulsions
      3. Phobias
V. Suicidality, Homidicality, and Impulse Control
VI. Insight and Judgment
VII. Reliability

Source: From LaBruzza (1994), pp. 105–106. Used with permission of the publisher.

Figure 17.1 Outline for the mental status interview

TERMS AND EXPRESSIONS USED FOR INTAKE, EVALUATIONS, ASSESSMENTS, AND REFERRALS

The following are some general terms that are commonly used.

PRIMARY LANGUAGE	When you see this on a form, give the person's native language and, if it is not English, tell how well the person functions with English.
PRESENTING PROBLEM	In one or two sentences, tell why the client is coming *now*. Use the person's own language.
PAST PSYCHIATRIC HISTORY	Use incomplete sentences. Give dates, approximately how long, and summarize if there is much detail.
FUNCTIONAL ABILITY	Note particularly if the person is able to carry out age- and stage-appropriate skills and tasks. Also note any recent change.
MOODS/EMOTIONS	What does the person or the person's family say?
PHYSIOLOGIC	What does the person or the family say about the person's appetite, sleep, and sexual activity?
THINKING	What is the person saying about how he or she is thinking? Are you able to follow the patient's thinking?
PERCEPTION	Are there any hallucinations?
ORIENTATION/COGNITION/MEMORY	Does the person think he or she can find his or her way? Does the person know where he or she is? Does the person remember well?
MENTAL STATUS EXAM	This is a word picture that tells what the person looks like *now*, not all the time.

The terms that follow are used in evaluating the various areas that need to be evaluated in determining a person's mental status. The numbers indicate the major categories you must cover in a mental status report.

I. General Description

A. Appearance

You may find it average, meticulous, slightly unkempt, disheveled, with body odor, no makeup, makeup skillfully applied, garish makeup.

1. Grooming

Meticulous - means it is too perfect, unusually so
Skillfully applied - means the person is made up to look like a model
Garish - means the person looks outlandish

SELF-NEGLECT

Always indicate when you think this is present. It would involve such things as having body odor or looking disheveled and unkempt. Dress would be dirty, stained, or rumpled.

Dress: You may find it casual, business, fashionable, unconventional, immaculate, neat, stained, dirty, rumpled.

Immaculate - means too neat
Unconventional - refers to clothes inappropriate to the setting
Fashionable - is fine unless the person looks like something out of *Vogue* in an office in a small town or average city

2. Physical Characteristics
Note those features that are outstanding. Look at body build, important physical features, handicaps. Note voice quality. Is it strong, weak, hoarse, halting?

3. Posture and Gait
Note gait, and any need for devices such as a cane or crutches. Look at coordination and gestures. For instance, does a right-handed person make most of his or her gestures with his or her left hand? Something like this could be a clue to neurological difficulties. Does the person limp, or appear to slump? Does the person seem unsteady or shuffle?

B. Attitude and Interpersonal Style

Look at the attitude the person has with you. You may find it cooperative, attentive, frank, playful, ingratiating, evasive, guarded, hostile, belligerent, contemptuous, seductive, demanding, sullen, passive, manipulative, complaining, suspicious, guarded, withdrawn, obsequious.

Seductive - too close a relationship too soon; might call you by your first name or touch you
Playful - jokes, uses puns, self-deprecating humor
Ingratiating - goes along with whatever you think; wants to please
Evasive - talks, but gives nothing
Guarded - is more reserved than evasive; contributes the bare minimum, often with suspicion
Sullen - angry and somewhat uncommunicative
Passive - barely cooperates, needs to be led; generally without overt hostility
Manipulative - asks for special favors, uses guilt, solicits pity, threatens
Contemptuous - superior, sneering, cynical

Demanding - sense of entitlement
Withdrawn - volunteers little, appears sad

HOSTILITY	*Always* note when the person is hostile.
UNCOOPERATIVE	*Always* note when the person does not or cannot cooperate.
INAPPROPRIATE BOUNDARIES	*Always* note if patient is too friendly, touches you, or attempts to draw you out personally.

Watch your own emotional reactions to the client. This will give you important clues.

Facial Expression: You may find it pleasant, happy, sad, perplexed, angry, tense, mobile, bland, flat.

Bland - intense material, but looks casual
Flat - no facial expression

C. Behavior and Psychomotor Activity

Look at the quality and quantity of the client's motor activity. You may find the patient is seated quietly, hyperactive, agitated, combative, clumsy, limp, rigid, or has retarded motor function. You may find he/she has mannerisms, tics, twitches, or stereotypes.

Seated quietly - uses normal gestures, but does not move around much
—*Hyperactive* - is busy with his or her hands, possibly feet
Agitated - cannot sit still (could be secondary to antipsychotic medication)
Combative - looks ready to hit, threatening
Awkward - inability to manage activity like sitting in the chair or writing; drops things (may be part of the illness or reaction to medication)
Rigid - sits like a tin soldier
Mannerisms - these are unconscious repetitive actions
Posturing - where the person assumes certain postures and holds them inappropriately
Tics and twitches - less voluntary body movements
Stereotypes - four mannerisms strung together

MOTOR HYPERACTIVITY	*Always* report this when you see a lot of hyperactivity, restlessness, and agitation. It may indicate a manic state, reaction to medication, or anxiety.
MOTOR RETARDATION	*Always* report this when you see the patient moves slowly, in a constricted manner with minimal motor responses. Speech and thought are slowed, often depressed. Depression can give appearance of cognitive impairment.
MANNERISMS AND POSTURING	*Always* indicate mannerisms you see and any posturing.
TENSION	*Always* note tension, particularly if the person seems tense and the interview does nothing to relax the person.

SEVERE AKATHISIA	*Always* note severe restlessness. (Sometimes it may be part of an illness, and sometimes it may be due to medication. If the physician believes it is due to an illness and increases the medication, the person may grow much worse. Therefore, try to establish when it started, how long it has gone on, and whether it has grown worse recently.)

Note the Following When Present: Pacing, fidgeting, nail biting, trembling or tremulousness (a common side-effect of lithium carbonate and tricyclic antidepressants), abnormal movements such as rocking or bouncing, grimacing (particularly strange facial movements)

TARDIVE DYSKINESIA	*Always* note this condition if you see it or suspect this is what you are seeing. It occurs among psychiatric patients who have been on antipsychotic medications over a long period of time. The term literally means "late appearing abnormal movements" and seems to involve the muscles of the face, mouth, and tongue. Sometimes the trunk and limbs are also affected.

These movements can be slow and irregular (athetosis) or quick and jerky (choreic). All the movements are brief and involuntary and purposeless. A person may twist the tongue and lips, make odd faces, bounce or tap the feet, or actually writhe and squirm in the seat.

CATATONIC BEHAVIOR	*Always* note this behavior. It is generally a sign of severe depression or schizophrenia, catatonic type. It generally appears as a rigidity of posture wherein attempts to reposition the person are rigidly resisted. The person may voluntarily pose in bizarre and inappropriate ways. In waxy flexibility, the limbs of the person will remain in the position in which they are placed.

There is also a catatonic excitement wherein the patient engages in almost continual, purposeless activity that is nearly impossible to interrupt. Sometimes the patient engages in echolalia (repetition of everything he or she hears) or mimics and imitates others during this episode.

D. Speech

Speech is important because it is the primary means of communicating. Important to note are such things as rate, clarity, pitch, volume, quality, quantity, impediments, use of words, the ability to get to the point, and articulation.

You may find speech to be a normal rate, slow, hesitant, rapid, pressured, monotonous, emotional, loud, whispered, mumbled, precise, slurred, accented, stuttering, stilted, rambling.

> *Pressured* - not rapid but constantly talking; cannot be interrupted (often a sign of a manic episode). Person appears to have racing thoughts.
> *Monotonous* - no variation in tone
> *Emotional* - very expressive
> *Accented* - note a native accent and also if the patient seems to accent certain words or syllables
> *Impoverished* - may say very little either because of depression or because he/she is being interviewed in a language other than his/her native one; may also indicate a lack of facility with language

NEOLOGISMS   *Always* note when the person makes up
                                       entirely new words with idiosyncratic
                                       meanings. (This can occur due to aphasia or
                                       brain injury due to accident or stroke.)

Neurological Language Disturbances: You should be able to identify these disturbances. Strokes, head trauma, and brain tumors can cause the patient to lose his or her facility with language. Try to determine if the client has always had a language difficulty. Patients with schizophrenia may use loose associations as they talk. Those in a manic state may be prone to flight of ideas.

> *Aphasia* - loss of ability to understand and produce language; damage usually to left hemisphere of the brain (left-handed people often have this in the right hemisphere)

Type and extent of aphasia depends on location and extent of brain injury. There is

> *global aphasia* - can neither speak nor understand, read, write, repeat words, or name objects
> *Broca's aphasia* - can understand written and spoken language, but has trouble expressing his/her own thoughts verbally
> *Wernicke's aphasia* - inability to understand language and uses fluent, bizarre, nonsensical speech (The person may also act strangely and appear euphoric or paranoid or agitated. It is easy to think this is a psychotic thought disorder, but in schizophrenia the person is generally able to write and speak in his/her language, repeat words, and name objects.)
> *Dysarthria* - difficulty articulating due to problems with the mechanisms that produce speech. This sometimes produces distorted or unintelligible speech. The person usually can read and write normally. Ask the patient to repeat "No ifs, ands, or buts" to hear dysarthria better.
> *Perseveration* - defined as the persistence "in repeating a verbal or motor response to a prior stimulus even when confronted with a new stimulus" (LaBruzza, 1994, p. 113). The client may give the same answer to different questions or stay on the same subject.
> *Stereotypy* - "constant repetition of speech or actions" (LaBruzza, 1994, p. 113). The patient may pull a shoe on and off, twist and untwist the hair, repeat the same phrase or word over and over. These behaviors appear to be ritualistic and are common in childhood autism.

Give verbatim examples of what the client has said to support your assessment of speech.

II. Emotions

A. Mood

This is the way a person is feeling at any given time. You may find it euthymic, depressed, sad, hopeless, empty, guilty, irritable, angry, enraged, terrified, expansive, euphoric, elated, sullen, dejected, and anxious. Ask yourself, what seems to be the dominant mood of the person?

> *Euthymic* - normal mood
> *Expansive* - feels very good and is getting better
> *Euphoric* - out-of-sight happy
> *Anxious* - worried and distressed

B. Affect

Affect refers to the underlying flow of moods. This would be the outward expression of the emotional state. You can see it in the way patients use and position their bodies and in their tone and manner of speaking. You may find it broad, appropriate, constricted, blunted, flat, labile, anhedonic.

> *Broad* - normal range of moods
> *Appropriate* - appropriate to the situation
> *Constricted* - restricted range of emotional expression
> *Blunted* - even more restricted
> *Flat* - no change of mood, unemotional
> *Labile* - rapid change in mood (crying, then laughing)
> *Anhedonic* - incapable of any pleasurable response, depressed

BLUNTED AFFECT	*Always* note a blunted affect where you find no change in mood throughout the interview and no change in facial expression. It generally indicates depression.
EMOTIONAL WITHDRAWAL	*Always* note if the person seems emotionally withdrawn to you. The person would be inexpressive and probably have a blunt affect.
EXCITEMENT	*Always* note if the person seems inappropriately excited to you. It means he/she is overly enthused or terrified about the given situation.
FULL RANGE OF AFFECT	This refers to an appropriate affective response to the entire interview. *Always* note inappropriate affect (such as giggling when there is nothing funny happening), as this can be a sign of schizophrenia.

C. Neuorvegetative Signs of Depression

In major depression, body functioning often becomes irregular. Always inquire about sleep and appetite, and report a loss or gain of more than 5 percent of body weight. Listen for symptoms such as changes in energy levels, interest, and enjoyment in everyday activities, sexual functioning, constipation, and weight changes (LaBruzza, 1994, p. 115).

Initial insomnia - trouble falling asleep

Middle insomnia - middle-of-the-night wakening

Terminal insomnia - early morning wakening. Depressed individuals will often wake several hours earlier than usual and feel most depressed in the morning.

Hypersomnia - Some depressed individuals, especially those with bipolar disorders, tend to sleep a great deal.

III. Cognitive Functioning

A number of medical and neurological problems, as well as substance abuse, will affect one's cognitive functioning. The concern is that many patients who have a disease of the brain may appear with what seems to be emotional and behavioral changes. In taking the history from the client, note previous levels of functioning or previous emotional problems. If these are appearing in middle or late life, it is quite possible the person has a brain disease.

A. Orientation and Level of Consciousness

Nearly all the clients who come to you will be alert and aware of their environment and their bodies. You may, however, occasionally see individuals who are inattentive, drowsy, or who have a clouded consciousness. If these symptoms are present, use the proper term to indicate the person's level of awareness and briefly describe how the patient exhibits this level. Medication can contribute to these stages as well.

Lethargy - the person has trouble remaining alert and appears to want to drift off to sleep, but can be aroused. The person has trouble concentrating on the interview and seems unable to maintain a coherent train of thought.

Obtundation - the person is difficult to arouse and needs constant stimulation to stay awake. The person may seem confused and unable to participate in the interview.

Stupor - the person is semicomatose, and it takes vigorous stimulation to arouse the person, who cannot arouse him/herself. There is no normal interaction during the interview as a result.

Coma - this is the most severe consciousness problem wherein the person cannot be aroused and he/she does not respond to any stimulation.

Oriented x3 - means the person is oriented as to who he/she is, where he/she is, and when it is. Even when a person is having difficulty with consciousness, he/she may be oriented. If orientation problems occur as a result of lack of consciousness, it typically happens that the sense of time is affected first, followed by the sense of place, and finally by the sense of person. To be fully oriented requires an intact memory; thus, disorientation means there are memory deficits.

Ask for current date - reasonably accurate dates are acceptable.

Ask where the person is - you can also ask for a home address, the present city or state, or for directions from here to the person's home or another familiar place. Sometimes people confused about place will behave as if they are at home or in another very familiar setting while in your office.

Ask who the person is - ask for personal identifying information (age, birth date, name). Ask if the person recognizes or knows other people who might be present. Does he/she know the relationship to these other people?

B. Attention and Concentration

Attention: Can the person remain focused on the interview?

Always note inability to pay attention and if the person appears easily distracted. If you feel a need to test this in the person, you can use digit repetition. Say five numbers, then ask the person to recite them back to you.

Concentration: the person can concentrate on one thing for an extended period of time. Concentration is needed to learn new tasks and for academic success and can be tested by asking the person to perform a complex mental task. (Serial 7s is one way of testing, by asking the person to add in increments of 7 or subtract from 100 by 7s. Be sure your instructions are on the client's level of education, and do not use this exercise if there are severe academic problems present. Be careful not to humiliate the person!)

C. Memory

Memory involves the ability to learn new material, to retain and store information, to acknowledge and register any sensory input, and to retrieve or recall stored material. When there are problems, they usually have to do with three areas:

1. Registration
2. Retention
3. Retrieval

Destruction of significant parts of the brain will cause problems with memory. All memory deficits should be noted. The physician or clinician will want to do further tests. If you suspect something, ask others who know the patient about their perceptions of the patient's memory functioning.

Short-term memory - refers to immediate recall limited to about seven items and generally lasts for about one minute. Some problems may be due to inattention, so evaluate attention before memory.

Long-term memory - rehearsal allows material in short-term memory to convert to long-term memory. Anxiety about the interview or the person's situation or even depression can interfere with this.

Amnesia - inability to remember

Anterograde amnesia - cannot learn *new* material

Retrograde amnesia - cannot recall *recent past* events

Head injuries - most common deficits are inability to recall names, recent events, spoken messages, forgetfulness, forgetting to do something important. The person may have trouble telling you what he/she is experiencing with his/her memory. Memory loss may be permanent if there was severe or repeated head injury.

Transient global amnesia - lasts minutes to several hours and is usually seen in older people. The person experiences sudden confusion, loss of memory, and disorientation and cannot recall what happened during the time period in question. Retrograde amnesia will be present. Person will be distraught, asking for reassurance as to where they are and what they are doing. This is caused by an insufficient amount of blood to the brain.

Memory Testing: First, ask the person if he or she has been having any problems with memory. A family member may be able to shed some light on memory issues if any exist. During the interview, note memory lapses and difficulty recalling what the interviewer has just said. If you notice memory loss, note it so that further testing can be done.

All the tests for memory that follow would only be done if you had considerable question about a person's memory.

To Test Immediate Recall: Use a random list of digits, saying them in a normal tone of voice, about one digit per second. Ask the person to repeat them. Start with two digits and keep adding until the person fails. Give the person two times to try this. If he or she fails at five digits or less, there is reason for concern about sustained effort, attention span, and immediate memory. Anxiety and depression are the most common reasons people fail this test (LaBruzza, 1994, p. 125).

To Test Recent Memory: Ask the person to recall events that have happened in the last few hours or days before he or she came to see you. You might ask what he/she had for lunch or where he/she parked the car. It is helpful if you can validate the answers with someone close to the client who knows. Another way to test is to ask about something that may have happened or been discussed earlier in the interview. With some people, you might give three or four unrelated words and ask them to recall these words after a short interval. Begin by saying the words in a normal tone of voice and ask them to repeat the words back to you. Note how many times the client must do this before learning the words. About three to five minutes later, ask the client to recall the words. With a normal memory, the person should be able to recall them (LaBruzza, 1994, p. 126).

To Test Remote Memory: You can ask the client about personal events in his/her life and commonly known public events that happened in years past, such as major news stories. Use material that should be known by a person who is reasonably well informed. If the person does not appear to be able to do this test because of a lack of education, a difference in culture, or because he or she is mentally retarded, decide carefully what you will ask the person (LaBruzza, 1994, p. 128).

D. Ability to Abstract and Generalize

Proverbs   Cultural background and intelligence can influence how well a person thinks abstractly or how well he/she can deal with similarities. Proverbs are generally used to see how well a person thinks abstractly, and one needs to have a general fund of information to be able to use proverbs in this way. Tell the person you are going to say a proverb and you would like the person to tell you in his/her own words what he/she thinks the proverb means. Then judge how concrete or abstract the reply is. Repeat the person's response verbatim in your report.

Individuals who are psychotic or on the verge of psychosis will often indicate this in their response to a proverb. Use proverbs that are free of gender and racial bias. The following are some proverbs you can use (LaBruzza, 1994, p. 129).

> A stitch in time saves nine.
> A rolling stone gathers no moss.
> Don't judge a book by its cover.
> Two wrongs don't make a right.

For this last proverb a person thinking abstractly might say, "Just because someone has done something wrong doesn't mean you can correct it by doing something wrong yourself." On the other hand a person whose thinking is concrete or literal would be more likely to say, "Well two wrong decisions don't equal one right decision."

Similarities and Differences   Ask the client to tell you how two objects or two events are different or alike. This will require the client to think somewhat abstractly about categories and relationships. Name two items and ask the client how these differ and how they are similar. The following are some combinations you might use (LaBruzza, 1994, p. 129).

> Apples and oranges
> Trees and flowers
> Houses and cars
> Dogs and cats

E. Information and Intelligence

To get an idea of the person's overall intelligence, ask questions that tap the person's fund of general information. It should be information known by the general public. Again, you must be sensitive to the person's cultural background, level of education, and intelligence. The following are examples of some questions you might ask (LaBruzza, 1994, p. 130).

> Who were the last four presidents?
> Who is the governor of the state?
> How many weeks are there in a year?
> What is the capital of the state (or the country, or France)?
> Who was Mark Twain?

IV. Thought and Perception

When a person's perceptions are disordered, it offers important clues to what the diagnosis might be. Here you want to know how the person actually perceives himself, the world around him, and others in his world. What does he think, and what thoughts and concepts are most on his mind? Perception is the way in which we form an awareness of our environment. When the person has difficulties with perceptions, he will often perceive his world inaccurately (LaBruzza, 1994, p. 131).

A. Disordered Perceptions

Illusions - the person either misperceives or misinterprets a sensory stimulus. A tree branch brushing the side of the house in the wind sounds like people entering the house, or a dishwasher running sounds like people talking in another room.
Hallucinations - in the absence of external stimuli, the person perceives something. The most common hallucination is hearing voices. Voices generally increase when the person is around white noise. If you can, find out who is talking, what they are saying, and how the person feels about it. Is there a command for the person to do something? If so, include the command in your report. Some commands are dangerous to the person or to others. *Always* note hallucinatory behavior.
Depersonalization - the person feels estranged or detached from herself
Derealization - the person feels detached from what is going on around her. Be sure to note this. A person who dissociates cannot always be sure that what is happening is real (LaBruzza, 1994, p. 132).

B. Thought Content

Distortions - a person distorts a part of reality. A woman with anorexia believes she is fat when she is thin. A person who is well believes his cough indicates tuberculosis. A person whose neighbor does not think to wave assumes the neighbor is angry.
Delusions - an inappropriate idea from which a person cannot be dissuaded using the normal means of argument or evidence. Sometimes it is culturally inappropriate as well. Evidence to the contrary has no effect. Always report the content of a delusion. Note if the delusion is incongruent with the client's mood. Delusions indicate psychosis. *Always* note if delusions are present.
Paranoid delusions - believes she is being singled out for harassment or is being controlled by forces outside of herself. She may have an entire system of interconnected ideas developed that support her delusion. Common to schizophrenia are:

Thought withdrawal - belief that one's thoughts are being taken out of one's mind by an outside force

Thought insertion - belief that thoughts are being placed into one's mind by an outside force

Thought broadcast - belief that thoughts are being taken and broadcast so that others know what one is thinking.

Always describe suspiciousness.

Grandiose delusions - the false belief that one is extremely important or a false belief that one is imbued with special powers. *Always* describe ideas of grandeur and any grandiose behavior.

Somatic delusions - false beliefs about one's physical health

Delusional guilt - falsely believing that one is the reason or cause for terrible things that have happened or will happen

Nihilistic delusions - a false belief in the meaninglessness of life and all events and circumstances, in nothingness; hopelessness; belief in the end of the world

Ideas of inference - refers to the ideas the person holds about what others do to affect him/her

Ideas of reference - refers to beliefs that people are talking and thinking about one. Messages on TV and radio are meant specifically for this person.

Magical thinking - means belief in astrology or a superstition, or the person thinks he has magical powers in his words, thoughts, or actions. This thinking is found in children who have not developed reality testing. It is part of human development and is not pathological until it becomes extreme, as in obsessive-compulsive disorder or a delusion. Always check religious beliefs or cultural background and see how this thinking fits with what is going on in the person's life and these aspects of the person's life (LaBruzza, 1994).

THOUGHT CONTENT	*Always* specify unusual or important thought content such as:

What the person is suspicious about
What the person feels guilty about
What the person is preoccupied about

BIZARRE BEHAVIOR	*Always* note any that you witness or any that is reported to you by reliable others.

C. Thought Processes

You may find the person's thought form to be spontaneous, logical, goal-directed, coherent, impoverished, blocking, nonspontaneous, incoherent, perseverative, circumstantial, tangential, illogical in form. You may find it to have loose associations or flight of ideas. You may find it contains neologisms, or it is distractible.

Flow of ideas - refers to the quality of the associations the person makes between ideas or between points in the person's discussion. Note the stream of the client's thoughts, the rate of thinking, the coherence, the continuity, and whether the thought process is goal-directed (LaBruzza, 1994, p. 134).

Spontaneous - means you do not have to keep asking questions. The person readily volunteers information.

Goal-directed - the person answers the main questions as to why he/she came, what he/she needs, and does not stray to other related topics.

Impoverished - the person uses words but is very skimpy with them. There are too few ideas, and thinking is slow. Often attributable to depression or schizophrenia.

Racing thoughts - the person thinks rapidly. Speech appears pressured. Often attributable to manic or hypomanic state.

Blocking - the person stops, pauses, and starts somewhere else. There is an interruption to the normal flow of speech. Person may appear to forget where they were in the conversation when they resume talking.

Circumstantial - the person appears to throw in too many irrelevant details. Client has too many ideas associated with one another and too many digressions. Often thought to be a defense against dealing with troubling issues or feelings (LaBruzza, 1994).

Perseverative - the person goes over and over the same point or idea.

Flight of ideas - the person goes from one thought to another in logical sequence, but he/she is headed far from the original topic.

Loose associations - the person's points do not hang together logically. Ideas shift in an apparently unrelated way. Characteristic of schizophrenia.

Illogical - what the person is saying does not make sense.

Incoherent - there is no meaning; the speech is disorganized; the person may be schizophrenic.

Neologism - the person makes up new words.

Distractible - person cannot stay focused; may indicate mania.

Clang association - "The sound of a word, rather than its meaning, triggers a new train of thought" (LaBruzza, 1994).

Tangentiality - means "veering off" on somewhat related, but irrelevant, topics. May show a difficulty with goal-directed thinking. Common in mania and hypomania (LaBruzza, 1994, p. 136).

Overvalued ideas - the idea might be possible, but it is used or seen by the person to explain more than it could possibly explain.

CONCEPTUAL DISORGANIZATION	*Always* note conceptual disorganization. This refers to an inability to conceptualize the problem clearly and may involve a number of the terms given previously, such as loose associations, flight of ideas, tangential thinking, or incoherent content.

D. Preoccupations

These are thoughts and issues that appear to be the primary focus of the patient's thinking. There is an "obsessive quality" to the preoccupation (LaBruzza, 1994, p. 136).

Somatic preoccupation - focus on bodily functions, physical health. There is a hypochondriachal quality to the preoccupation. List and describe somatic concerns. Do not assume these are not real problems without proper medical documentation.

Obsessions - persistent thoughts that are intrusive and unwanted and that appear to haunt the person (LaBruzza, 1994, p. 136). The person may hold an idea that is not true in that intensity.

Compulsions - actions that are often the "counterpart of the obsession." These are "persistent, intrusive and unwanted urges" to take some action. If one does not complete the action, there is intense anxiety. These actions can be repetitive and ritualistic, such as checking the stove, counting steps, and straightening picture frames (LaBruzza, 1994, p. 136).

Phobias - these are "irrational, intense, persistent fears" of such items as dogs, heights, elevators, insects, leaving home, closed spaces, and flying (LaBruzza, 1994, p. 136). The person will go to great lengths to avoid the situation or object of the phobia.

V. Suicidality, Homocidality, and Impulse Control

Suicidality and Homocidality

You have a clinical and a legal responsibility to assess whether the person is a danger to him/herself or to others. A person who is dangerously impulsive or holds thoughts of suicide or homicide would need special observation. When conducting an interview, look for these thoughts. These are called suicidal ideation or homicidal ideation. Discover if there are plans to carry out these ideas or if the client seems to have an undeniable intention. Thoughts present some danger, plans make the situation more dangerous, and the clear intention to go forward with the plans creates an extremely dangerous situation.

The mildest form would be thoughts, followed by actually making plans. If the client tells you she has purchased a gun and learned the intended victim's work schedule, you may assume there is a plan. If the client tells you he has saved three months worth of medication for an overdose, this would appear to be a plan. If the client has a plan and expresses to you his or her obvious intention to carry out this plan, the situation becomes more serious (LaBruzza, 1994).

Always include the plan, the means, and the timetable in your report. If the client is homicidal, include toward whom the thoughts are directed as well.

Impulse Control

When you assess impulse control, you want to know how the person can deal with aggressive urges, sexual urges, or strong desires to carry forward any plan not particularly well considered. Look at ways the client has handled stressful situations in the past. Is there a history of acting on impulse without much thought as to the consequences? Does the person seem unable to tolerate stress? How did this person handle stress in the past? Is there a history of uncontrolled aggressive behavior, either sexual behavior or hostile behavior? Can this client tolerate frustration?

The three behaviors to look for in childhood would be setting fires, cruelty to animals, and bedwetting. When these behaviors all occur in childhood, they appear to be significantly associated with cruel adult behavior. In adulthood, you might find punching holes in the wall, smashing furniture, slitting his or her own wrists, drinking excessively, or turning to drugs or a drug overdose.

VI. Insight and Judgment

Insight - The person understands that he/she is "suffering from an illness" or has emotional or personal problem. Make a note if the client completely denies any problems or denies having any part in a problem that obviously affects the client's relationships with significant others. People hold insight in "varying degrees" and may show only partial understanding of their difficulty (LaBruzza, 1994, p. 137).

Judgment - The person can critically evaluate his/her situation and make good decisions about a course of action. Look for risky behavior in the past that could have been potentially harmful (practicing unsafe sex, binge-drinking and driving, and so forth). "Assess whether clients are able to understand the potential consequences in their behavior" and can plan preventive measures. You might ask what the client would do if he/she spotted a fire in a movie theater or what he/she would do if he/she found a stamped sealed and addressed envelope (LaBruzza, 1994, p. 138).

VII. Reliability (Accuracy of the Client's Report)

You need to state briefly your impression of the client's reliability and accuracy in giving you the details of his/her situation. If the person is psychotic, the material presented is likely to be extremely unreliable. A person who is suffering from dementia or delirium may be having considerable difficulty remembering what has happened or what is happening now. Some clients deliberately tell falsehoods to qualify for disability or to give false impressions of themselves to the worker (LaBruzza, 1994, p. 139).

VIII. The Environment

Sometimes you will be asked to go to someone's home to do an assessment or to do an interview. A person's surroundings often hold clues to the way a person is currently structuring his or her life.

Inappropriate surroundings - means the person has arranged furnishings inappropriately. It may be that furniture blocks doors and windows or that the windows are covered oddly, perhaps with tin foil or some other material. There might be strange wires leading nowhere, odd decorations, or strings of odd things hung across a room. You might find household objects being used inappropriately. Sometimes a person believes the people on TV can see into his/her home or the people on the radio can hear what he/she says. In those cases, you may find these devices covered or blocked in some way.

Be very careful in making these judgments. At one time a worker decided that windows partially covered with foil-wrapped insulation were a sign of something amiss in his client. It turned out the man was doing that to save energy. In another home a worker decided the blanket over the TV indicated her client was paranoid. In fact, it turned out the TV was very new and the client was very proud of it; she had put a blanket over it the day before when her young nieces came to visit so the TV would not get scratched. It helps to inquire matter-of-factly about what you see.

Waste and trash: Look at the way the person keeps his/her home. Are there unusual collections of junk or trash? Are there piles and piles of paper and magazines everywhere, or collections of string, bags, and other objects? Sometimes the home is very cluttered or is very dirty, with unwashed bed linens and with unwashed dishes stacked all over the kitchen. This tells you something about the person's capacity to attend to the routine details of living, or it may indicate a debilitating mental illness. You might find urine or feces on the floor or walls. You might find the pets in the home neglected, their food bowls full of rotten food. Sometimes the house might be overrun with strays the person has taken in, but is unable to adequately care for.

Note: It is not unusual for a middle-class worker to assume that the manifestations of poverty are the signs of a person who is mentally ill. You must be very careful not to ascribe mental problems to someone who is actually too poor to live by middle-class standards.

EXERCISES USING THE MSE VOCABULARY

Instructions: See if you can fill in the blanks for each of these questions. This exercise is simply to acquaint you with words you might encounter in the course of your work.

1. The chart on Mr. Kling reads that he has conceptual disorganization. What was the psychologist referring to? *Thought Processes*

2. In a staff meeting, the psychologist and the case manager discuss the fact that Mrs. Purdy seems to have racing thoughts. They are talking about what? *Thought Processes*

3. The neurologist's report comes for Mr. Engler. The diagnosis is Broca's aphasia. What does that mean? *Can understand written and spoken language but have trouble expressing verbally his thoughts.*

4. When the psychiatrist saw Mrs. Nguyen, he said her mood was euthymic. He meant *Normal Mood*

5. Mr. Kissel's speech is described as impoverished. That means his speech is *lacking of facility with language*

6. Another term for severe restlessness is *severe akathisia*

7. Another way to describe a broad range of moods is *full range of affects*

8. When Dr. McCoy said Mr. Perkins used neologisms, Dr. McCoy was referring to *speech*

9. Mrs. Dell has been on antipsychotic mediation for some time, and now she shows signs of Tardive dyskenesia. That means she *have late abnormal movements.*

10. When Mrs. Jones was described by the psychologist as seductive, he meant that Mrs. Jones was *having to close of a relationship too soon.*

11. The chart says that during the interview with Mr. Landon at the prison, where he is staying after being arrested for drug possession, Mr. Landon's speech was guarded. That means his speech was *reserved*

12. When the ambulance crew called in about 93-year-old Mr. Keller, they said they were bringing him into the hospital, but he was oriented times 3. They meant *he knows who he is, where he is and when it is.*

13. Mrs. Harris complained of trouble falling asleep. In the chart this was written as *initial insomina*

14. Discussing Mr. Rodriquez's delusions with you, the psychiatrist remarks that Mr. Rodriquez has thought withdrawal. The psychiatrist means *he thinks his thoughts are taken out of his mind.*

15. The psychologist is describing his interview with Mrs. Carter. "She tends to perseverate," he noted. He means that Mrs. Carter *is repeating a verbal or motor response to a prior stimulus with a new stimulus,*

16. Mr. Trong has been depressed ever since he came to this country from Vietnam. Lately the depression has worsened. Today his therapist calls from the International Center asking if you can arrange hospitalization for Mr. Trong because he is catatonic. Mr. Trong is *severely depressed.*

17. Miss Aller is homeless and mentally ill. The mission calls to say that Miss Aller believes that her thoughts are being taken out of her head and broadcast so that others know what she is thinking. Another way to write this is _thought Broadcast_

18. When Mr. Cruz talks about the accident, he tells you he feels detached from himself. Another word for that feeling is _Depersonalization_

19. The chart tells you that after the accident, Mr. Cruz had retrograde amnesia for awhile. That means he _cannot Recall Recent past events_

20. The record that comes from the hospital on Ms. James states that she suffered from terminal insomnia during her stay there. That means that she _has Early morning wakening_.

21. After the stroke, Mr. Torres was described as having Wernicke's aphasia. That means _inability to understand language._

22. During the interview, Miss Bell constantly pulled her gloves on and off, on and off, in a ritualistic fashion. One word for this kind of behavior is _Compulsive_

23. Miss Bell was asked to interpret several proverbs to test her _Ability to think Abstractly._

24. Mr. Lincoln was seen with his wife for marital difficulties the couple was having. He said he personally had no problems at home and if his wife was saying there were problems at home, these were her problems and she was bringing these on herself. What might we say about Mr. Lincoln's insight? _idea of inference_

25. The family doctor of Mrs. Fong calls to arrange for an appointment. She believes Mrs. Fong is depressed and "is showing strong neurovegetative signs of depression." What does the family doctor mean by that? _Body functions are irregular_

26. When you interview Mr. Marks you notice that the *sound* of a word seems to trigger a new line of thought for Mr. Marks. You call this _clang Association_

Chapter 18

Receiving and Releasing Information

SENDING FOR INFORMATION

You have done an assessment of the client who is seeking treatment or services through your case management unit. Before you can adequately plan, it may be a good idea to obtain records or summaries of past treatment or services the client might have received. This information gives you a clearer picture of the client and will help you to plan for that person. Knowing what has been done in the past will prevent your trying programs and services with which the client has had little success. During the assessment, have the client sign the appropriate forms; and immediately after the interview, mail or fax those forms to the other agency so the information can be obtained in a timely manner.

IF YOU RELEASE INFORMATION

Sometimes clients will bring you a form and ask you to release their records to another agency. It is important to protect clients when releasing information. Even though they tell you it is all right to do so, clients may not fully realize the consequences of giving out information. It is your responsibility to see that clients understand what could happen when information is released.

For example, a woman with mild mental retardation had suffered from depression in the past. She was hired in a position at the police station as a cleaning person. When the police personnel office learned of her involvement with MH/MR, they asked for her records. She was very willing to oblige. Her case manager talked to her, however, and the two of them decided to release only a brief, noncommittal summary with few of the details.

The facts of a client's situation belong to the client. These facts are not yours, even though they are in your possession. Releasing them without written consent is unethical—and also illegal, unless covered by the exceptions we looked at in the chapter on ethics. Before working on this chapter, you would do well to review the section in the ethics chapter on confidentiality.

You will find when you are practicing case management that some places have a blanket form. The client signs the form for a number of places listed on the form for an unspecified length of time. You may have encountered this at your doctor's office where you may have been asked to sign that your records can be released to a specialist if you are referred to one. These forms, while saving the staff time, are of questionable legality. You may have to observe this practice in an agency where you work, but we will not be using that format here—nor do we recommend this, due to liability and confidentiality considerations.

DIRECTIONS FOR USING THE RELEASE FORM

The form you will use is in the Appendix. It will allow you to obtain information you need to plan for the client. Here is information you will need to correctly fill out the form.

1. RE: refers to the client. Neatly and clearly write in the client's name on that line.
2. DOB: refers to the client's date of birth.
3. "I hereby authorize" is the client speaking. Place your name on that line, as you are the one who needs and will be receiving the material or information.
4. On the line that reads "to receive information about services rendered to me from," place the name of the agency where the client received or is receiving services.
5. State the name of the primary practitioner at the other agency, if you have this information (Dr. Wentworth, Miss Jones). If there was more than one person, list all their names. (For instance, the client may have seen Dr. Wentworth for medication and Miss Jones as his therapist.)
6. The statement that the information will be used "for the purpose of _____" refers to why the material is being sought. Generally, filling in the blank with "goal planning" or "treatment planning" or "developing a service plan" will be sufficient. Sometimes a doctor with whom you are working is seeking the information for "ascertaining past medications for the purpose of prescribing." Do not go on a fishing expedition, asking for material that you think would be interesting to read, but which bears little impact on what you are planning for the client.
7. Dates of service covered should be very specific. Do not use open-ended statements (for example, "during the summer of 1994" or "last year"). The dates in this portion should read something like: "August 1 – August 30, 1995" or "December 22, 1993 to April 24, 1994" or "August 1995 through September 1996." If you and the client are not sure about dates, estimate them as close as you can. You probably cannot be quite as specific in that case.
8. Information you are seeking is always best obtained in summaries. You do not want everything because you cannot store it all. Only seek what is relevant. For instance, do not ask for school records if the person is an adult with no school problems. Do not seek all the hospital records if the discharge summary will suffice.
9. The authorized signature is the signature of an adult or the signature of the parent or guardian if the client is a child. You countersign the form, as well, and date it.

EXAMPLE OF A RELEASE FORM

A release form signed by a woman who has been in a drug and alcohol treatment program in another state is shown in Figure 18.1. The case manager needs to understand what took place there in order to plan adequately for the client.

RELEASE OF INFORMATION

RE: *Marcella Renova*

[name of client]

DOB *3/7/40*

[client's date of birth]

TO WHOM IT MAY CONCERN:

I hereby authorize *Carlos Sabatini*

[name of case manager]

at *Wildwood Case Management Unit*

to receive information about services rendered to me from

Spring Garden Drug and Alcohol Clinic, 123 First St., Anytown, PA

[name of agency]

Primary practitioner was *Kathryn M. Parsons*

This information will be used for the purpose of *Treatment Planning*

Such information may be transmitted under the conditions stated below, and/or as required by State or Federal statute or order of the court. This release is effective for a period of 90 days from the date of this release.

Dates of service covered in this request *4/97 through 6/99*

Information to be given may include:

[ ]	medical records	[ ]	social/developmental
[x]	discharge summary	[x]	psychiatric evaluation
[x]	psychological evaluation	[ ]	educational record
[ ]	vocational evaluation/summary	[ ]	treatment summary
[ ]	other _____		

I have read this form carefully and I understand what it means.

Marcella Renova	*4/8/98*
authorized signature	date
Kathryn M. Parsons	*4/8/98*
staff signature	date

To the agency or professional person receiving information with this release:

THIS INFORMATION HAS BEEN DISCLOSED FROM RECORDS WHOSE CONFIDENTIALITY IS PROTECTED BY STATE LAW. STATE REGULATIONS PROHIBIT YOU FROM MAKING ANY FURTHER DISCLOSURE OF THIS INFORMATION WITHOUT PRIOR WRITTEN CONSENT OF THE PERSON TO WHOM IT PERTAINS.
THIS CONSENT TO RELEASE OF INFORMATION CAN BE REVOKED AT THE WRITTEN REQUEST OF THE PERSON WHO GAVE THE CONSENT.

Figure 18.1 Sample release of information

WHEN THE MATERIAL IS RECEIVED

When you have received the material, you have the elements to construct a fairly accurate clinical picture of the person who has come to you for help. You have the person's initial description of the problem, followed by a more detailed explanation and a thorough social and medical history. You have sought and noted the client's expectations for services and done some preliminary planning with the client. You have at least a general idea of how the *DSM-IV* categories might or might not apply to your client. You have assessed the person's mental status and received information to support or enhance your own conclusions. It is time to structure a service plan for your client.

RELEASE OF INFORMATION EXERCISES

Exercise One   Intake of a Middle-Aged Adult

Instructions: Using the blank form in the Appendix entitled "Release of Information," send for information relevant to planning for one of the cases you are developing. Choose the information for which you are sending carefully, and be able to explain how this information will assist you in planning for your client.

Exercise Two   Intake of a Child

Instructions: Using the blank form in the Appendix entitled "Release of Information," send for information relevant to planning for the child's case you are developing. Choose the information for which you are sending carefully, and be able to explain how this information will assist you in planning for your client.

Exercise Three   Intake of a Frail, Older Person

Instructions: Using the blank form in the Appendix entitled "Release of Information," send for information relevant to planning for one of the cases you are developing. Choose the information for which you are sending carefully, and be able to explain how this information will assist you in planning for your client.

Exercise Four   Maintaining Your Charts

Instructions: Place your completed release forms into the appropriate clients' charts. They should be placed under the long assessment forms in the charts. If there is more than one form, staple them together.

Developing a Plan with the Client

Chapter 19

Developing a Service Plan

INTRODUCTION

The next step is to develop the service plan for your client. Following is the step-by-step process for doing that. Remember that you will not be giving these services. Instead, you will be determining what the person needs, based on your evaluation or assessment, and based on your discussions with the client about services he or she feels are important. The services and agencies you choose for your clients will be those that will help clients attain goals that you and they see as important for them.

There are several types of goals for clients. You might develop treatment goals to treat conditions such as mental illness, emotional problems, or drug or alcohol abuse. Treatment goals generally refer to mental health programs, drug or alcohol rehabilitation, or physical health interventions. Not all of your clients will need treatment goals. Some clients will have problems involving poverty, home arrangements, and material needs. These are general living goals for clients that will improve their situation, perhaps by providing appropriate in-home care, better transportation, or a more appropriate living arrangement.

USING THE ASSESSMENT

You will receive an evaluation or assessment for each person in your caseload. In this case management unit, you will have done the assessment or evaluation yourself, which is commonly how it is done. The evaluation should tell you what the person's major problems are and what areas of the person's life need attention. Generally these problems are the reason the person is seeking help or is being referred to a therapeutic or supportive program. The evaluation will help you to determine what goals you think you should work on with your client. Figure 19.1 lists common goals to consider for clients. You may think of others.

These goals are very broad, long-term, and general. Choose those goals that fit your client's most outstanding needs and give them more specificity. For example, your client might need "to make positive changes regarding stressors and coping mechanisms." You would not word it in just this way. You would give it more specificity by writing something like this: "Client needs job with better work hours" or "Client needs help with marital problems."

To obtain (or maintain) housing	To obtain (or to maintain) food and clothing
To make positive changes regarding relationships with family members	To acquire (or to improve) daily living skills
To obtain (or maintain) mental health treatment or rehabilitation services	To obtain (or to maintain) drug and alcohol treatment or rehabilitation services
To obtain (or maintain) physical health treatment or rehabilitation services	To make positive changes regarding stressors and coping mechanisms
To recognize precipitants to hospitalization and take appropriate action	To obtain legal assistance on legal issues
To obtain (or maintain) social support	To obtain (or maintain) transportation
To obtain (or maintain) recreation and leisure time activities	To obtain (or maintain) an education program
To obtain (or maintain) employment	To obtain assistance with financial issues
To obtain a new role in life	To develop realistic goals for the future
To continue living independently	

Figure 19.1   Goals to consider for clients

Later when the client is referred to an agency for help in meeting the goals, the agency will set up smaller, more specific goals and a plan for the client to meet those goals.

CREATING THE TREATMENT OR GOAL PLAN

When you go to the treatment planning conference or service or disposition planning meeting, you should have a provisional plan for your client. After this meeting, your next step is to create the final service plan for your client. In putting together the provisional plan, first, write out the goals so you have a clear understanding of what you and the client see as important in this case. Then, using the "Treatment or Goal Plan" form found in the Appendix, follow the steps below.

1. Place the client's name and next of kin on the top line of the form.
2. Check the box for an initial plan, as you have never done one before on this client.
3. Fill in the date the plan was created. Make the review date three months from this date, unless in your judgment a review should take place sooner. For instance, if the person is in need of inpatient drug and alcohol services and those services are only expected to last ten days, you would want to check on the service sooner.
4. Indicate who helped to formulate this plan (the client, daughter, mother, father).
5. For the level of case management, wait until you are in the planning meeting to decide that with others. You may have an idea about the level when you go to this meeting.
6. Print your name as the case manager.

7. The box containing the material on the *DSM-IV* should be left for completion in the planning meeting. Again, you can have a rough idea of what you will suggest for these axes, but wait to make that final until you have heard from others.
8. Sign and date the form at the bottom, placing your name on the line for the case manager. Your instructor will sign and date the supervisor's line.

To fill in the boxes:

1. For each category on the left
   a. Circle or highlight "Strength" if this is an area of strength for the client—that is, if there are no problems for this client in this area or if the client brings resources to this area. You can make a notation or comment by way of explanation in this box.
   b. Circle or highlight "Need" if there is a problem in this area that you intend to address. You can make a notation or comment by way of explanation in this box.
   c. *Do not circle either STRENGTH or NEED if the area in the box is not applicable to the client's problem.*
2. Write the goals in the boxes underneath the heading "Goal(s)." You can base the goals on the list shown in Figure 19.1, or you can come up with others.
3. In the boxes in the last column, indicate the name of the agency to which you will refer the client to meet the goals. Try to get as much service from one agency as possible. Sending the client to five or six different agencies can be confusing. For clients who have many needs, choose agencies that give more comprehensive services.

Note

Not every client who comes for help has a mental health problem. A person who has run out of fuel oil might have a mental health problem, but if the reason that person is seeking help today is to get a voucher from you for more fuel oil, the mental health problem is not relevant. Be careful, therefore, not to assume that everyone you see must also be mentally or emotionally ill. Look at the V codes in the back of the *DSM-IV* for problems people often encounter in life that might become the focus of clinical attention, without meaning that the person is mentally ill. For many people who are depending on the social service setting, even these will be irrelevant.

Later in this chapter (see Figure 19.2) is an example of a goal plan for a client.

INVOLVING THE CLIENT AND THE FAMILY

Only ten or twenty years ago clients would request help from an agency and be told by that agency what help they would receive. Clients had little or no input into these decisions, and clients who objected or had other ideas were often viewed as difficult and uncooperative. Agencies developed services in part based on the skills and expertise of the employees at those agencies. They might offer services that were badly needed in their community, and they might offer services that did not meet the needs of any but a select group of people. Clients were expected to fit into the services offered and were often labeled untreatable if they did not.

Today, all that has changed. Several factors have fostered these changes. When the state mental hospitals began to shrink their patient populations, patients were brought back to their communities and back in contact with their families and other support systems. What soon became obvious was that many families were unable to cope with the mental illness of their family member without support, and without that support the patients were often abandoned by their families. Agencies began to recognize the value of including families in planning and implementing treatment. In more recent years, children have been receiving services in ever-greater numbers. This has involved families in planning and implementing the treatment or service plans of their children.

Now agencies are expected to engage both the client and the client's family, when appropriate, in developing plans and support that will enable the identified client to live successfully in his or her community and will give the family the support they need to remain an intact family. Individuals who suffer from chronic mental illness do much better in their communities if their families have not cut them off. Children can take better advantage of services if the entire family is considered in the planning. For these reasons, human service professionals are careful to listen to what the client wants to receive from services and to what the family may need to support the client.

An inappropriate use of family involvement would occur if a worker ignored the wishes of an adult who came seeking services for emotional or situational problems and did not desire family involvement, or if a worker involved family members after a client asked specifically that his/her family not be involved in any way. In these situations, you follow the wishes of the client.

INDIVIDUALIZED PLANNING

All of this has led to individualized planning for clients. Clients no longer have to fit into a certain program or go without treatment. Instead, case managers develop plans for their clients that are specific to that person. No two plans should be exactly alike.

An example of poor planning can be found in the caseload of Alicia, whose clients were all over 70. Alicia was only 24, having recently graduated from school, and to her people over 60 seemed elderly. In planning for her clients, Alicia assumed they were all in the same need of supervision and support. She referred all of them to Homemakers, Inc. for household help. She insisted, when she could, that each of them attend the local senior citizen center. In addition, she tried to line up Meals-on-Wheels for each of her clients.

The clients began to talk among themselves about Alicia's lack of appreciation for what they needed in their lives right then. Marguerita did need Meals-on-Wheels, but only temporarily while her broken arm healed. She found going to the senior center a chore because she wanted to stay home and read and watch her favorite shows on TV with her neighbor. Delbert belonged to a retired businessmen's club and wanted to go there in the afternoon. He found it annoying that Alicia called to check on him when he failed to go to the senior center. Delbert had come in for help with portable oxygen. When Leslie refused to use the Homemaker, Inc. services, Alicia went to Leslie's home to find out why. Alicia appeared astonished that Leslie was able to keep her own home and often entertained family and friends.

Alicia is an example of a case manager who does not view her clients as individual people. She lumped them all together in a category of "elderly people." She devised her service plans based on her assumptions about what elderly people should need.

UNDERSTANDING BARRIERS

Barriers are those circumstances that prevent you from understanding the client fully or prevent the client from taking advantage of services and even your support. If you are aware that barriers will diminish the effectiveness of your plan, you can modify the plan accordingly. This helps you to plan in a way that prevents problems later and helps the client to take full advantage of the plan. It is a two-way street. Barriers prevent workers from fully understanding and helping the client, and they prevent the client from being able to take full advantage of the plan being developed. Even when you have sought and addressed the barriers you have been able to identify, you may refer your client to a place where barriers will appear again. Care in planning is the best way to make sure your client can truly take advantage of the service or treatment.

Here are some common barriers:

Language	The client may not be able to communicate adequately with others because of a difference in primary language. The worker may not be able to communicate with the client because the worker does not speak the client's language.
Culture	The client may be unable to negotiate an unfamiliar culture. The worker may not understand the client's culture or may be inclined to judge the client's culture by the worker's own cultural standards.
Disability	The client may not be able to handle all the details of the plan. The worker may overestimate or underestimate the extent of the client's disability.
Resources	The client may lack the resources to fully participate in the plan (for example, transportation or clothing suitable for a job interview). The worker may see the client's poverty as a barrier to the plan or fail to take the lack of resources into account.
Mental illness	The client may be unable to communicate clearly or follow through with the plan. The worker may be afraid of the illness or may fail to understand how the illness impacts the client's capabilities.
Mental retardation	The client may be unable to communicate clearly or follow through with the plan. The worker may see the client as a child or may fail to understand how the disability impacts the client's capabilities.

SAMPLE GOAL PLAN

Figure 19.2 shows the goal plan for a 34-year-old man, Larry McCune, who was in a severe car accident four years ago. He has come into the case management unit requesting help with symptoms he originally thought would disappear or that he could handle

TREATMENT OR GOAL PLAN

CLIENT _Larry T. McCune_   Next of Kin _Lydia McCune (Mother)_

Initial plan [ ]   Updated plan [ ]   Date _3/6/01_   Review Date _6/6/01_

Developed with _Larry McCune_

Level of case management _Resource Coordinator_   Case Manager _Kathy Torres_

Provisional DX: Axis I _309.81 PTSD, Chronic with delayed onset_   Axis II _V71.09_

Axis III _Headaches following head injury_   Axis IV _Marital breakup_   Axis V _50_

TYPE	STRENGTH/NEED	GOAL(S)	REFERRAL
INCOME/FINANCIAL SITUATION	(STRENGTH) / NEED		*Receives income from disability policy*
HOUSING LIVING ARRANGEMENT	(STRENGTH) / NEED		*Owns home*
VOCATIONAL	(STRENGTH) / NEED		*Would like to return to job*
EDUCATIONAL	(STRENGTH) / NEED		*Has engineering degree*
TRANSPORTATION	STRENGTH / (NEED)	*To be able to drive independently*	*Problems with transportation can be addressed in therapy*
MEDICAL	STRENGTH / (NEED)	*To have headaches evaluated*	*Micael C. DeFilippo Neurologist*
ACTIVITIES OF DAILY LIVING	STRENGTH / NEED		
LEGAL	STRENGTH / NEED		
RECREATION & LEISURE TIME	STRENGTH / NEED		
MENTAL HEALTH	STRENGTH / (NEED)	*To reduce or eliminate symptoms of PTSD*	*Linden Counseling Center*

DRUG & ALCOHOL	STRENGTH		
	NEED		
FAMILY RELATIONSHIPS &/OR SOCIAL SUPPORTS	STRENGTH	*To reestablish ties with 9-year-old son*	*Linden Counseling Center*
	(NEED)		
OTHER	STRENGTH		
	NEED		

Kathy Torres	*3/6/01*	*Marshal Potts*	*3/6/01*
Case Manager Signature	Date	Supervisor's Signature	Date

Figure 19.2 Sample treatment or goal plan

on his own. The police report states that he was driving his car with his family in the car. He approached a busy intersection and failed to stop for a yellow light that turned red while he was in the intersection. He was hit and two people were killed, one of them his 4-year-old daughter. Larry is well educated, with several degrees, and works as an engineer in a position he enjoys. Since the accident, however, he has been unable to work consistently and has had numerous arguments with his subordinates. His wife left a year ago, taking the remaining child with her, a 9-year-old boy. Larry is complaining of severe headaches and is not clear if these are from stress or the accident in which he suffered what was diagnosed as a mild head injury. He is asking for help because of recurring nightmares that involve the accident and an intense fear of riding in cars, which recently has spread to using any form of public transportation. He states he feels detached from other people and lately has thought that perhaps his life is really over.

Note that the case manager has put together the plan so that the client can receive his services almost entirely from one source, in this case the Linden Counseling Center. Personnel at that center will address this client's specific needs. A neurological assessment will be done by a neurologist, whose findings will ultimately be taken into account at the Linden Counseling Center in working with Mr. McCune.

BROAD GOAL PLANNING EXERCISES
Exercise One   Planning for a Middle-Aged Adult

Instructions: Using one of the blank "Treatment or Goal Plan" forms found in the back of your book, develop a plan that addresses the immediate or most important needs of the client you are following. Look at both strengths and weaknesses.

This will be the tentative plan that you will take to the planning meeting. Make certain that you have addressed those areas of your client's life that are most troublesome right now to the client. If one of your clients is facing a poverty situation, do not assign a *DSM-IV* diagnosis.

Exercise Two   Planning for a Child

Instructions: Using one of the blank "Treatment or Goal Plan" forms found in the back of your book, develop a plan that addresses the immediate or most important needs of the child you are following. Look at the strengths and weaknesses of both the child and the child's family.

This will be the tentative plan that you will take to the planning meeting. Make certain that you have addressed those areas of the child's life that are most troublesome right now to the child and his or her family with regard to this child.

Exercise Three   Planning for an Infirm, Older Person

Instructions: Using one of the blank "Treatment or Goal Plan" forms found in the back of your book, develop a plan that addresses the immediate or most important needs of the older person you are following. Look at both strengths, including natural supports the person might be able to call on, and weaknesses.

This will be the tentative plan that you will take to the planning meeting. Make certain that you have addressed those areas of your client's life that are most troublesome right now to the client.

Exercise Four   Maintaining Your Charts

Instructions: The "Treatment or Goal Plan" form should be placed under your release forms. This is a tentative form and does not need signatures. You will clean up and revise your recommendations in the planning conference, and then put the revised form in the chart in place of this original.

Be sure that every chart you are following has a plan in it. Charts without plans are often the reason funding and accreditation sources withdraw support.

Exercise Five   Checking Services

Instructions: If you have a question about a service in your community that you believe might provide the services your hypothetical client needs, call that agency for information. Most agencies have brochures or other informational literature they will send to you.

Chapter 20

Preparing for Service Planning Conference or Disposition Planning Meeting

INTRODUCTION

After you have completed your assessment on each new client and done a tentative plan, your agency might hold a meeting in which plans are made for the client's care or services. In some agencies this is done informally. In small agencies, particularly, individual case managers may make those decisions by themselves, referring the clients to other services in systems that will have more formal case management.

In some places, children who come into the system are presented by their case manager to a "children's panel" consisting of child psychologists, child psychiatrists, social workers, pediatricians, and others who serve children. Many places are adopting this same format for serving clients from different populations; in this situation, the case manager presents the case to representatives of any number of agencies serving or specializing in that population. Together the group decides what combination of services would best suit clients in their current situation and gives a diagnosis, if appropriate.

Where a person has both a drug and alcohol (D&A) problem and a mental health problem, if the agencies that address these two problems are not combined, representatives from them should meet together to decide what should be done. In the past a client could be turned down for mental health services because he was drinking and turned down for D&A services because he was suicidal. That kind of turf warfare at the expense of the client is no longer tolerated by funding sources that expect the client to be served.

In these meetings, decisions regarding the service your client will receive are made with others who have experience and come, perhaps, from different disciplines. When the meeting is over, a formal plan will be drawn up.

WHAT YOU WILL NEED TO BRING TO THE MEETING

You should consider bringing three items to these planning meetings.

1. *Tentative service plan.* You have already developed a tentative service plan for the client. Bring this tentative plan to the service planning conference.
2. *Human service directory.* As you work within the same social service system, you will come to know, without consulting a directory, which agencies are reliable

and which services are used most often by your agency in referring clients. As you begin your career, you need to know what human service organizations are available in your community. If there is a directory, bring that to the meeting so that you can work with your peers to find the best placement for your client.

3. *DSM-IV Handbook.* The *DSM-IV* is a large volume containing considerable information. If you are working in an area that is likely to use the *DSM* to give diagnoses, you might consider purchasing the *DSM-IV Handbook* that contains only the most basic information and is easier to carry with you when you go to meetings of this sort. Bring your *DSM-IV Handbook* to assist in making the provisional diagnosis.

GOAL ONE FOR THE MEETING: DIAGNOSIS

If this meeting regards a client seeking services from mental health/mental retardation or from drug and alcohol services, you will need to give a provisional diagnosis. Older people who appear to have some type of dementia would also receive diagnoses. Diagnoses in these service delivery systems are usually required for payment purposes and can be changed after the client is in the system and has been thoroughly observed.

GOAL TWO FOR THE MEETING: LEVEL OF CASE MANAGEMENT

In addition to the provisional diagnosis, you may assign the level of case management if your agency uses different levels. For our purposes, we will use three levels. The levels are as follows:

1. *Administrative case management:* where clients are placed in a pool with other clients who require little service or follow-up beyond the original referral. The client tends to function independently.
2. *Resource coordination:* where clients are often in need of services and assistance on issues like housing, medication, or therapy, but they generally do well with the services offered and do not pose a risk to themselves or others.
3. *Intensive case management:* where the client is someone at high risk for rehospitalization or for behavior that poses a danger to the client or other people.

GOAL THREE FOR THE MEETING: SERVICES

In every case management situation, the most important part of the intake process is to make decisions about the service the client will receive. Your assessment prepares you to make those decisions wisely. Remember that you do not give the service. Your task is to look at the material in your assessment interview, such as:

- The strengths and weaknesses of the client
- What the client said he or she expects
- What services this person could use well or fit into well

- What the major problems or presenting problems were (presenting problems are those that brought the person into the agency in the first place)
- Goals you have for the client and the client's stated goals
- Any other pertinent circumstance or information about the client that you feel is relevant

Using the material you have assembled, you (and presumably a team) will develop in this meeting a plan for services, and for treatment if needed, that matches the client's problems and expectations.

PREPARING TO PRESENT YOUR CASE

In the planning meeting, you will give the pertinent details that will help the group make decisions about where the client will be referred and the type of treatment the client will need.

Your presentation will be a short, oral summary of your case to the group. It should be given in an orderly manner. Before the meeting, review the details of the case. Bring the intake and assessment material with you, and refer to it if you do not know the answer to a question. In general, however, you should be able to answer most questions without reference to the material.

The elements of your presentation would include:

1. Why the client came to the agency. What were the presenting problems?
2. How the client presented in the assessment interview.
3. What did the client indicate were his or her goals or expectations?
4. Additional relevant information that would have bearing on the disposition of this case.

1. *Why the client came to the agency. What were the presenting problems?* Discuss why the client came to the agency. What were the outstanding problems he or she talked about during the phone intake and at the assessment interview? Mention any other outstanding problems that you feel should be addressed.

2. *How the client presented in the assessment interview.* Talk about any unusual or bizarre behavior. Describe any hallucinations or delusions the client might have had. Most clients will not have any of these. Be sure to talk about the client's affect, motivation, and insight.

3. *What did the client indicate were his or her goals or expectations?* What did the client say he or she wanted to have happen as a result of seeking services? Tell why the person sought help and what the client expects the outcome will be. Even though you will formulate the ultimate goal plan for this person, be sure to indicate the client's input here. Have a tentative goal plan ready, but expect others to see and suggest additional goals or to suggest changes.

4. *Additional relevant information that would have bearing on the disposition of this case.* If there is other information that the team should have in order to make a decision, be sure to discuss that. In addition, describe any unusual characteristics that might give a clearer picture of the client. Perhaps it is a quote the client made. Perhaps the client has worked for several years at the Humane Society and has four dogs he rescued. During the interview, he talks about his dogs and shows you pictures of them. This fact may not directly influence the treatment plan created to deal with his depression, but it will give the team a clearer picture of the client and the client's interests.

MAKING THE PRESENTATION

By following the format previously discussed, you will make a good presentation. Be prepared to talk about the case to the team for about three to five minutes. You should not have to read through notes or shuffle papers. Simply describe your client. After you are finished, the team will ask you questions about your client. You should be able to answer these without reference to your notes, but that may not always be so. If you do not have the information, turn to your notes when you need to.

COLLABORATION

Treatment or service planning, when it is done with a team, is a collaborative activity. Some case managers can become defensive as others suggest changes or additions to the tentative plan. One case manager was "upset" when the tentative diagnosis she had in mind was questioned by the psychiatrist. Feeling that a change in the diagnosis would reflect her incompetence, she resisted any change and became angry.

Another case manager was sure that the man whose situation he was presenting was not experiencing medical problems as a result of his drinking. He became sarcastic in the meeting when others asked that a medical examination be part of the plan. "I think I've seen enough alcoholics to know when someone needs medical care and when they don't," he sneered at the team.

In a third situation, a woman presented a case of a client suffering from severe anxiety. Another case manager questioned the diagnosis, stating it seemed more like posttraumatic stress disorder. The client's symptoms and complaints needed further evaluation by the team, and a discussion ensued. The case manager was asked questions as the team attempted to assemble more details in order to make the diagnosis. The case manager answered these questions, but she did so in a petulant manner and later reported that her "feelings were hurt" because "people I thought were my friends just ganged up on me in there."

On the other hand, it is important to ask questions of your colleagues in a collaborative and respectful manner. Do not grill the people presenting cases. Do not be dismissive of their interpretations of the situations or demean the conclusions the presenters have made. After all, the presenter is the person who actually saw the client, and his or her observations are extremely relevant to the decisions made on the client's behalf.

CONCLUSION

After the treatment planning conference or disposition meeting, you will write up a formal service plan for your client using the "Treatment or Goal Plan" form. When your goal plan is completed, you will:

1. *Meet with the client and discuss the plan*. Your first case note should be your first meeting with the client after developing the goals and referral options and having those confirmed in the treatment planning conference or disposition meeting. In this meeting, you will go over the plan with the client, or the client's parents in the case of a child, and note the client's response and any changes you make to the plan as a result.
2. *Make referrals* for your client to the agencies that will actually carry out the services.

PLANNING EXERCISES

Exercise One   Developing a Service Directory

Instructions: Before you go to a planning meeting, you need to know what services are available in your community. Gather information on various social service agencies. Some regional phone books contain special pages of social services. Some counties publish a directory you may be able to purchase. The library may have lists of agencies you can copy, or a local agency may have compiled a directory.

Be sure the directory you compile or obtain is relevant to the area in which you intend to work following completion of your courses. In this way, you will begin now to become familiar with what services are available.

Exercise Two   A Simulated Planning Meeting

Instructions: To simulate a planning meeting, form groups of no more than five students (otherwise it will take too long for everyone to present a case). Present one of the cases you have developed to the group following the instructions for presentation provided in this chapter. Be sure to bring the information on the client and the material you will need to plan services and give a diagnosis.

You can have your instructor act as the senior specialist who can confirm with you the proper diagnosis, or you and the team can arrive at your own decision.

Sign the revised planning form, and have your instructor also sign it, as your supervisor.

Chapter 21

Making the Referral and Assembling the Record

INTRODUCTION

After the service planning conference or disposition meeting, you developed a formal service plan for your client. You discussed this final plan with your client, and now you will refer the client to the agency or agencies that will carry out the treatment or service. Use the "Referral" form in the Appendix to make referrals to these other agencies.

Referrals are generally faxed or sent by e-mail to the agency to save time, although they can be sent by mail.

1. All referrals are coming from the case management unit for which you work, in this case the Wildwood Case Management Unit.
2. Write the name of the agency to which you are referring the client after the word "To."
3. Note the date the referral was made.
4. Write the client's name, address, and phone number after "Re" in the box.
5. Write the goals the referral is to address in list form.
6. The "Target Date" is the date you expect the goals to be met.
7. After "Review Date," place the date on which you intend to review this case plan to see if the plan is working.
8. Write your name in the blank for case manager.

DETERMINING DATES

The Target Date

When you set a target date, you are stating clearly to the providers how long you expect it will take for the service to actually obtain the goals and objectives you have for your client. You are, however, clarifying something else. You are informing the provider how long you will allow the service to be given without getting the desired result. If the target date is reached and the objectives have not been accomplished, it is time to stop this intervention and seek a more useful approach. When setting a target date, decide how long you are willing to continue to try this approach without seeing any result. Beyond that date, you will not continue funding, and at that point you will evaluate other options for the client.

Routinely physicians expect to see results from medications they prescribe for their clients. They know that if the medication has had little or no effect in a specified number of weeks, it is time to switch to another medication. Neither you nor the physician in this example can afford to administer a particular treatment plan for as long as it takes, no matter how long that might be.

The target date is influenced by two factors:

1. *Funding.* The amount of money available to spend on the service will influence how long the client can stay in the service. If you are using public funds or insurance, there may be a cap on the amount of money you can pay for a particular service.
2. *Goal.* The goal is a factor in the length of time needed. Some goals are short-term, such as a six-day detoxification program, while others are a substitute or a prevention for inpatient hospitalization and may require weeks or months of service.

The Review Date

The review date is the date you expect to review the plan to see if the plan is actually working. In most cases, plans are reviewed at least every 90 days. Therefore, a person in a partial hospitalization program for six months would be reviewed in three months. As noted, however, some services are short-term. In the case of a six-day detoxification service, you might make a quick review on the third day.

The purpose of the review date is to make sure that the client is getting what the client needs and what the referral stipulated and to be sure that if the plan is not working, no further money will be spent on trying to make it succeed.

SAMPLE REFERRAL FORM

Figure 21.1 contains a simplified version of what you are likely to see in a referral form in actual practice. Many provider agencies have their own referral forms they want the case management unit to use when making a referral to that provider agency. These forms ask for more information about the client, usually in a detailed summary. For our purposes, we will use a simplified referral form here.

The referral in Figure 21.1 is for Paul Bittinger, a person with chronic schizophrenia. He has been in and out of hospitals for acute episodes of hallucinations during which he believes the voices are giving him the power to walk on water. These voices, and his subsequent delusion that he is omnipotent and can walk on the surface of a nearby river, have caused him to jump in the river at times when it was high or there were huge, swiftly moving ice chunks. In an effort to help him manage his own illness and medications better, the case manager is referring him to a partial hospitalization program.

THE FACE SHEET

Your clients are now in your agency's system, and a record is assembled from the various forms and contacts you have had with these clients and those concerned with them. All charts on clients who have entered the system have a face sheet. (See the sample of a

REFERRAL

From: *Wildwood Case Management Unit*

To: _Grandon River Partial Hospitalization_
           [name of agency]

Date: _____ _10/13/01_ _____

Re: ___*Paul J. Bittinger*___
                    [name of client]

Address _____

_____

Hme Phone _____ Wk Phone _____

For the purpose of: [list goals] _____

_____ *Learning to manage medications* _____

_____ *Learning to manage the symptoms of his illness* ____

_____

_____

_____

_____

TARGET DATE ___*4/17/02*___   REVIEW DATE ___*1/15/02*___

CASE MANAGER ___*Phyllis Martinez*___
                            [signature]

Figure 21.1   Sample referral

completed face sheet in Figure 21.2.) This sheet lies on top of all the other information and contains essential information for anyone who might need access to it quickly.

You will find a blank face sheet in the Appendix. To fill out the face sheet:

1. Place the name of the client, the agency number you have assigned the client, and the client's address and phone numbers at the top of the form. (For our purposes here, give the client any agency number you wish. It should be at least four digits.)
2. If the person is a child, someone with a severe mental handicap, or an elderly person in need of a guardian, put the guardian's information in the next section. Most clients will not need a guardian, in which case you can leave this area blank or write N/A in the space.
3. Nearly everyone has someone they wish to have notified in case of an emergency. It may not be a blood relative; a friend or a neighbor is acceptable. The person the client indicates he or she feels closest to should go in the section under next of

kin. Indicate the relationship of that person to the client (for example, mother, sister, or friend) in parentheses.

4. Below the double line, from "Physician" on down, fill in only the information that applies. Not all clients will be on medication or have a physician involved in their case.

   a. If the client has a physician who is part of the service plan, record the physician's name. Otherwise, leave that space blank.

   b. If the person is on medication, fill in that information. Otherwise, leave those lines blank.

   c. If the person was given a diagnosis from the *DSM-IV* or has an overriding medical diagnosis, place that on the face sheet. Otherwise, leave the diagnosis line blank.

   d. Do list the place where most of the service plan will be carried out.

   e. Indicate the name of the person most responsible at that facility for seeing that the service plan is followed.

   f. List the main goals you developed for your client.

   g. If other agencies are involved, indicate that. It is not necessary, however, to put down every single agency that is involved with your client. Just note the primary ones.

   h. You are the case manager. Putting your name in the case manager space will indicate to your instructor that this is your chart or file.

   i. The "as of" date should be the date the service plan was completed in treatment planning conference. The date you will do your first review is entered on the form as well.

When you place the face sheet in the front of the chart, you have an organized collection of documents. Using a manila file folder, place the other forms on your client in this order, with the face sheet on the top and the referrals on the bottom:

1. Face sheet
2. Inquiry and referral form
3. Verification letter
4. Assessment form
5. Release of information forms
6. Service plan
7. Referrals

Now you can add to the chart all further contacts, letters, and monitoring activities that take place for this person.

EXERCISES TO ASSEMBLE THE RECORD

Instructions: Complete the following exercises.

1. Look at the completed forms on the clients you have developed. Fill out referral forms for each agency to which you intend to send the client for services. In each client's chart, clip these together and place them in the client's chart. Use a separate referral form for each agency.

2. Next, develop a face sheet for the front of each chart.

FACE SHEET
Wildwood Case Management Unit

Name _Lucinda Harris_ Agency # _2487_

Address _____

_____

Phone [Hme] _____ [Wk] _____

Guardian _N/A_

Address _____

_____

Phone [Hme] _____ [Wk] _____

Next of Kin _Jamal Harris (Husband)_
[If different from Guardian]

Address _____

_____

Phone [Hme] _____ [Wk] _____

Physician _Curtis B. Quimper, M.D._

Medications _Desipramine 75 mg. Q.D._

_____

_____

_____

Diagnosis _296.33 Major Depressive Disorder Recurrent_

Primary Treatment Facility _Linden Counseling Center_

Therapist _Michael Cedaneo_

Current Goals _To be able to return to work_

_____

Other Agency Involvement _____

Case Manager _Larry K. Imball_

As of _4/10/02_ Next Review Date _6/10/02_

Figure 21.2 Sample face sheet

3. Now assemble each chart as indicated in this chapter, using a manila file folder and placing the forms on your client in the following order, with the face sheet on the top and the referrals on the bottom:

   1. Face sheet
   2. Inquiry and referral form
   3. Verification letter
   4. Assessment form
   5. Release of information forms
   6. Service plan
   7. Referrals

Once you have written them, your case notes will follow the referrals when they are put into the chart.

Section 6

Monitoring Services and Following the Client

Chapter 22

Monitoring the Services or Treatment

INTRODUCTION

Some agencies do a considerable amount of case management. Such an agency will be responsible for large numbers of clients who fall within a certain category of need or difficulty. In these agencies, case management is broad and encompasses many services rendered to many clients. Usually county agencies, mandated to exist in each county by state law, operate in this way. Services for aging, for mental health and mental retardation, and for children and youth are all examples of this type of large case management organization. Here every case that is accepted into the system is managed by the central office. Sometimes the agency will elect to provide the services as well, using its own staff, rather than send the client to another service provider. In those situations, the case managers and those who provide the service work for the same organization. For instance, a county Office of Aging might contract out to another organization the management of its senior citizen centers, or it might decide to employ the managers and run the centers itself. This decision is usually based on cost-effectiveness or on the availability of services in a community.

Other agencies do very limited case management or only manage cases within their own organization, monitoring for success and failure of their own service. Often these are agencies that receive limited public money and must account for it carefully. An example might be a small partial hospitalization program for people with chronic mental illness. This program might be receiving referrals from a number of sources, some of them private physicians and some of them public agencies for mental health. Within the partial hospitalization program, clients might be assigned a person who functions as the case manager. This person would see that the plan is implemented and followed and would make corrections to the plan as needed. Agencies like this are generally accountable to a case management organization elsewhere that authorizes payment of the bill for that service and monitors the clients' progress toward their goals.

In another instance, a small agency may make only a few referrals to other services and must make sure the client is receiving the correct service and moving toward the goals within its own program. A client of a domestic violence program, for example, might be referred to legal assistance, a job readiness program, and public housing. A worker might act as the case manager, developing an individualized plan, implementing it, and monitoring the plan within the domestic violence agency and also among the few referral sources to help the client move toward her particular goals.

PURPOSE OF MONITORING

Many of the clients with whom we deal are receiving services paid for by public funds. These funds are limited and must be spent wisely. As the case manager, you are responsible for deciding what services a client will receive, based on the goals you have developed with the client. Then you must estimate how long it will take to reach these goals.

You might have a client with developmental disabilities who needs to develop a marketable skill. You refer the client to Goodwill Industries. You authorize services there for one year. This means that you have allowed for payment to be made to Goodwill Industries for one year of job training for this particular client. The expectation is that at the end of the year, the client will have a marketable skill.

You would then go to Goodwill Industries at regular intervals to see if the goal is being approached. Is the client moving toward a marketable skill? Are there obstacles and problems you did not foresee or that are new in the client's life? You cannot just refer the client to Goodwill Industries and leave him there for the year without ever checking up on how well things are going.

Angelica provides another example. She needed a supportive environment in order to be able to live with minimal supervision in the community. She had been hospitalized four times for severe depression, and it was only during the last hospitalization that the medical staff was able to combine medications for her in such a way that she really felt free of her depression symptoms. The severity of her depression and the number of acute episodes had prevented her from working or from living on her own successfully. Now, as the depression cleared, the staff was looking for a program where Angelica could begin to work toward an independent living arrangement.

The case manager in a mental health case management unit placed her in a small program called the Blue House, so named for the color of the house. There, three women lived for a period of 90 days, preparing to move out on their own. The time in the house was spent in normal activities, such as cleaning, cooking meals, decorating for the holidays, and going to the movies. During the day, the residents engaged in activities to help them secure a job and a place to live.

Case management chose the Blue House program specifically for Angelica, believing this was the best place for her to receive the support she needed to move toward independence. Her improvements and her accomplishments were reported to the case manager; and as Angelica began to make plans to move to her own apartment, the case manager collaborated in the planning, coming to the Blue House for meetings with Angelica and the staff.

Let us look at another example. Suppose you have a client who came into your transitional housing program after suffering years of marital difficulties and homelessness. You would develop a service plan for this client that addresses his most urgent needs. Perhaps you and he decide one of his needs is for a community college degree in order to be self-sufficient. In addition, he may need professional help in parenting his children, who appear to be out of his control. His children may need academic and recreational programs as well, and there may be a medical problem requiring attention. After you develop the plan, you would refer the client to the appropriate services. In this case, your program would not pay the actual costs of each of these services. A Pell grant and other aid will pay for college, the children's programs are supported by community funds and are free to the participants, and the parenting help is obtained through the public child protection agency's parenting classes.

Nevertheless, your agency is receiving public money to develop good service plans for clients like this one in order to move them from dependence to independence. The funding source expects you to develop plans that are usually successful. Too many poor plans and resulting failures might cause you to lose your funding. The plan you devise, therefore, must be in the best interests of the client. Thus, for these reasons, you carefully monitor your client's participation in the services and his progress toward the goals. Very often, in a situation like this, you would talk more directly with the client about his progress or obstacles rather than talking to the providers of the service. For instance, it would be an invasion of the client's privacy to go to the community college to check on his grades, but you would want to keep in touch with him so you know about problems that might arise and might cause him to fail if they are not addressed.

COLLABORATION

Collaborating with other agencies is not always as simple as it seems. There are times when people at other agencies begin to think they know the client better than the case manager and to develop goals and objectives without collaboration with the case manager. Sometimes they discharge clients or switch them to other services within their agency without telling the case manager.

You may also run into situations where an agency is willing to take any client referred to them, but gives very poor care or minimal service. While they appear to be an ideal place to put clients for whom there is no other service, this may not be in the best interests of the client.

In these situations, it is tempting to become angry with the agency or with specific individuals working there. Anger and unpleasant disputes can permanently spoil relations between entire groups of professional people, curtailing the system's effectiveness and creating a hardship for clients. There are other, better ways to handle disagreements.

Try going to the other professionals and expressing your concern. If that does not work or the same difficulties continue to occur with other clients, ask your supervisor to talk to their supervisors, or ask the head of your agency to work with the head of the other agencies, to create a positive agreement on how these issues will be handled in the future.

For your part, make every effort to be responsive to the concerns of the provider agency. Be available for planning meetings and reviews. Support decisions the agency wants to make that really are in the best interests of the client. In this way, all the professionals involved with the client present a united approach to the client's problems. If your client is creating a problem at the provider agency, go and help them develop a plan to deal with it, talk to your client if that will help, and invite the staff there to join you in looking for the causes of the problem.

Do not blame the staff in the provider agency—not to them personally or behind their backs. Refrain from gossiping about how poor the program is or the staff members are. Instead, use the communication skills you have perfected to raise your concerns and listen to the responses. By following these guidelines, you will serve your clients better and maintain the smooth operation of the delivery system.

WHAT IS MONITORING?

Monitoring is an ongoing review of clients' participation in the services to which they were referred. This review must be documented on paper in the client's file and will be part of the case notes of each file. Reviews are carried out by doing the following:

1. Talking to the clients regularly to see if they feel they are making progress and to learn if they are satisfied with the services. If clients point out needed revisions in the plan, this must be in your notes, as well as what was done in response.
2. Contacting the person in the agency or program who is primarily responsible for the clients' meeting their goals in your service plan. What does this person think of the clients' progress toward completion of their goals? Should the service be continued or discontinued? Should the service be modified in any way?
3. Contacting other people, agencies, and services involved with the client. For instance, if the client is a child, contact the parents. If the child is in school, contact the teacher to learn if there are noticeable changes.

LEAVE THE OFFICE

Too many case managers want to monitor their clients' progress from the comfort of their offices. They find it too much trouble to drive around town, find parking, go out in all kinds of weather, or visit in homes located in areas that make the case manager uncomfortable. Talking to providers and clients on the phone and insisting that clients and their families always come to you is both arrogant and somewhat lazy.

When your client is admitted to the hospital, visit her there. When your client begins a job-readiness training program, drop in to check with him there. When your client is in a program to gain independent living, stop by to see how she is doing.

Clients live and work and strive toward their goals in a community. Only by visiting them in the settings where they are living and functioning can you get a true picture of who your client is and what the real obstacles or accomplishments are. The telephone is certainly useful as a support to your monitoring efforts, but relying on that exclusively cuts off important information and opportunities to strengthen rapport.

CONCLUSION

The funding source, the client, and the client's family all expect you to collaborate to improve some aspect of the person's life. They expect you to put into place new circumstances and skills to prevent future problems, and to catch wasteful or inappropriate planning before too much time and money have been invested. Skillful monitoring can help you accomplish all of this.

 **Chapter 23**

Documentation and Recording

DOCUMENTING CLIENT ENCOUNTERS

Once a person is in the agency's system and is receiving services, it is your responsibility to keep a record of all contacts relevant to this case. These contacts will be with the client or with those connected to the client in some way, such as providers of service, family, and counselors. You will document these contacts on the form entitled "Contact Notes" found in the Appendix. Sometimes clients will come to your office because something has happened in their lives that is upsetting to them or because they need something. At other times, clients might call on the phone because of a problem or need. You might see clients at the agency where the client is receiving services when you go out to make a site visit. Every contact of this sort will be documented.

The reason for keeping accurate records and documenting all contacts with or related to the client is primarily for legal and administrative purposes. Legally you need to be able to show that service is being given to the client for which you are being paid. Administratively you need a record that documents the activities on behalf of the client and all contacts related to the client so that case managers are not relying on memory to reconstruct what has happened before.

These notes should focus on your client, and not on you. The treatment plan begins to go out of date soon after it is written, due to the changes in programs and clients' lives. The purpose of your notes is to keep the record current.

WRITING CONTACT NOTES

Your contact note in the chart should *always* include:

1. The focus of the interview
2. Your assessment based on a concise summary of behavior, appearance, affect
3. Any resolution that takes place
4. The reason for the next contact or the follow-up that will occur

See the examples of contact notes in Figures 23.1 and 23.2. Figure 23.1 identifies the four parts of the contact note. See if you can identify the four parts in the example shown in Figure 23.2.

A case note of this type is never more than six or seven sentences. Write concisely in order to facilitate others having access to information. No one can sit down and read through lengthy, descriptive narratives. Your notes must be clear, and they must be concise.

LABELING THE CONTACT

On the lefthand margin for every case note, place:

1. The date
2. The type of contact in parentheses (Office Visit, Phone, Group, Program or Site Visit, Collateral Contact)

For example:

> 4/3/96 (Office Visit)
> 9/8/96 (Phone)
> 12/6/96 (Group)

Types of Contacts

Types of contacts include the following:

Collateral	A collateral contact is with someone other than the client such as the client's mother, minister, or nurse. Be careful in all collateral contacts that you have the client's permission to make that contact and, if the client is a child, that you have the parent's permission to talk with this person.
Office Visit	The client was seen in the office.
Phone	The client called you on the phone./You called the client.
Site Visit	You went to the site where services are being given to your client and evaluated those services or discussed problems that have arisen.
Group	You saw the client in a group and you are noting what took place during that contact. Sometimes clients are seen in groups to use time more efficiently.
Home Visit	The client was seen at home.

DOCUMENTING SERVICE MONITORING

In addition to your direct contact with the client, document your efforts to monitor the delivery of service to your client. Those notes should be set up the same way and labeled:

> 2/14/96 (Site Visit) . . . If you went to the program the client is enrolled in and talked with both the client and the staff or attended a treatment planning meeting at the other agency

Contact note broken into four parts

Focus of the Interview
Linda came into the office today to discuss her medication.
Your Assessment
She appeared somewhat disheveled and tearful and indicated her belief that the medicine is "not strong enough."
The Resolution
An appointment was set up for her to see Dr. Wentworth on July 2, She was advised to remain in her program where her depression can be closely moni-tored.
Reason for the Next Contact
Client will return July 2 after her visit with the doctor to let CM know what was done about her medications.

Figure 23.1

Contact note. Can you identify the four parts?

Mark called today asking for a voucher for public transportation in order to get to Polyclinic Medical Center for kidney dialysis. Suggested he go by county transportation for more direct service. Mark seemed bright and to be pleased with the results of dialysis. He will call next week to confirm that county transportation has begun to pick him up.

Figure 23.2

that focused on your client's treatment or services from the program. Be sure to name the agency in your notes.

2/24/96 (Collateral Contact) . . . If you spoke to the contact person at the other agency by phone regarding your client's progress or services.

For example:

> 1/6/00 (Site Visit) Met with client and social worker at Riverview Center to monitor client's progress toward job readiness. Client seemed eager to begin job search and somewhat annoyed at the length of time the job-readiness classes are taking. It was agreed that client will begin his job search with support next week, while completing the remaining segments of the training. Client will notify CM in two weeks regarding outcome of job search.

All your case notes must reflect the fact that you made efforts to monitor the services delivered to your client.

DOCUMENTATION: THE FINISHING TOUCHES

There are a number of ways to make your case notes sound professional. The following sections contain some tips for writing better notes.

Hostility

Do not use your notes to release hostility. When we are angry with someone, it is easy to sound sarcastic, facetious, or even annoyed with the person. Make sure your notes do not reflect any negative feelings you might have about another person.

Interactions

Your interaction with the client may be the most important thing that occurred. This could be a verbal exchange or some other form of interaction. Document what happened and what you observed.

Documenting Significant Aspects of Contact

Some aspects of the contact with a client are extremely significant. These are clues to the client's state of mind or situation. When you think it is significant, document the following:

- The client's appearance
- The client's dress
- The client's facial expression
- The client's mannerisms
- The client's response to others or to activities
- The client's participation concerning interaction with you or participation in the services referred to
- The client's attitudes concerning interaction with you or participation in the services referred to

Clarity

One factor that can make your notes more professional is your precision and clarity. Many people write vague notes or notes that give only general descriptions. Be clear about what you are documenting. Do not use vague terms or indefinite statements. For example:

Poor: *Alice was friendly today.*	Better: *Alice initiated the conversation, joking about her Christmas shopping.*
Poor: *Bill was upset today.*	Better: *Bill was concerned about the possibility that he could lose his job.*
Poor: *Marcella got along well in the program today.*	Better: *Marcella's affect was improved today, and she participated in preparing lunch for the group and in both group sessions.*

Quotations

What the client has said to you may be extremely important. You may feel that this information should go into the record. The rule is to place *only* the client's exact words in quotation marks. If you paraphrase what the client has said, do not use quotation marks.

Contradictions

Your progress note must not contradict other previous notes without explanation. If the plan is changed, that must be documented. If the client regresses or improves considerably, this should be documented. There should not be gaps that lead the reader to conclude that something happened that has not been documented.

Language

You have learned a rather extensive vocabulary that is used by professionals in the field. To the client, these words can sound like jargon. Sometimes new workers use a lot of jargon in an attempt to sound knowledgeable. Avoid jargon. Write your notes in language the client or the child's family can understand.

Disabilities

When writing about a person with a disability, be sure to use language that accurately reflects the person's life circumstances and does not label the person in pejorative ways. Here are some guidelines from the Three Rivers Center for Independent Living in Pittsburgh:*

First person	Identify the person first, rather than the disability. Use *person with disability* or *a person who is deaf* rather than *disabled person* or *deaf person.*
Disability	The terms *afflicted with, suffering from, cripple,* and *victim* are all unacceptable. They emotionalize and sensationalize, often to induce pity. The term *handicapped* is based on the image of a person with a disability on the street with a cap in his/her hand, begging for money. Except when citing laws, regulations, or environmental conditions, such as stairs (for example, the stairs are a handicap to her), always use *disability.*

---

* Adapted with permission from the Three Rivers Center for Independent Living, 7110 Penn Avenue, Pittsburgh, PA 15208-2434.

Wheelchair	People are not confined to their wheelchairs; they use them for mobility. Say he or she *uses a wheelchair*, not *wheelchair-bound* or *confined to a wheelchair*.
Blind	This term refers to total loss of vision. *Partial vision*, *partial sight*, or *visual impairment* are more accurate terms in some cases.
Deaf	This term refers to total loss of hearing. *Partial hearing*, *hard of hearing*, or *hearing impairment* are more accurate terms in some cases.
Nonverbal	*A person who cannot speak* is preferred over terms like *mute*, *deaf-mute*, or *deaf-dumb*. These terms imply the person is also deaf or also unintelligent. The inability to speak does not indicate intelligence.
Congenital disability	This is a disability that has existed since birth. Do not use the term *birth defect*. *Defect* is derogatory and is not a synonym for disability.
Learning disability	This term refers to a disorder affecting the understanding or use of spoken and/or written language.
Mental disorder	This term describes any of the recognized forms of mental illness or other emotional disorders. Terms such as *neurotic*, *psychotic*, or *schizophrenic* are pejorative labels.

Be careful about using phrases such as *he overcame his disability* or *in spite of her handicap*. These terms inaccurately reflect the barriers people with disabilities face. They do not succeed in spite of their disabilities as much as they *succeed in spite of an inaccessible environment or a discriminatory society*. They do not overcome their disabilities so much as they *overcome prejudice*.

GOVERNMENT REQUIREMENTS

States and the federal government have requirements for documentation that they will insist you follow in order to be reimbursed for your services to the client. These may vary from state to state and from one type of service to another. These funding sources treat the record as one would a blank check. Whiteout, erasures, and blank spaces are not acceptable to them. The following are some of the general rules that are often required:

1. Use black ink. Blue ink does not copy well.
2. All notes must be legible.
3. The client's name must be identified on each page. Sometimes you can use an agency number instead. For children, a date of birth on each page is often required. Do not use nicknames or initials.
4. When recording, place the actual date of the contact note in the margin.
5. Sign (do not initial) every note.
6. After your signature, add the date the note was written. It should be on or as close to the date of service as possible.

7. If the client is in ongoing service, every note must end with the next scheduled service date.
8. To correct a mistake in a note:

   Draw a line through the error—whether it is a letter, word, phrase, or entire paragraph.
   Write the word *error* above the line.
   Write the correction next to the word *error*.
   Sign or initial the line.
   Date your signature or initials.

9. If there are any blank lines left on a page, draw diagonal lines through the blank space.

DO NOT BE JUDGMENTAL

Do not write notes that sound as if you are sitting in judgment of the client. You may be inclined to judge some aspect of your client's life or behavior in negative terms, but that is not helpful; and when negative judgments are obvious in your case notes, they leave a legacy that can follow the client. Avoid judgmental words in your notes. Figure 23.3 contains a list of words that tend to sound judgmental and a substitute word that is more objective for each. Become familiar with these.

How to avoid sounding judgmental

Poor Words for Documentation (Judgmental)	Better Words for Documentation (More Objective)
dirty	unclean habits, poor hygiene
nasty	unpleasant
lazy	inactive
stubborn	resistive
nervous	anxious
wild	restless
bad-mouthing	argumentative
sarcastic	critical
mean	unpleasant, insulting others
troublesome	uncooperative
whining	complained of
glum	sullen
just sat there	passive
jittery	restless
foolish	used poor judgment
slow	had trouble completing
pushy	persistent
aggravating	irritated others

Figure 23.3

KNOW THE DIFFERENCE BETWEEN FACTS AND IMPRESSIONS

A fact is something you observed. Impressions are the clues you pick up from the client. Use words such as the following to introduce your impressions:

> Client seemed . . .
> Client appeared . . .
> Staff felt . . .

For example:

Poor: *Mary acted pleasant but was putting on a front.*

Poor: *Manuel wasn't telling the truth when he said he was comfortable with the new group.*

Better: *Mary was pleasant, but CM felt she was feeling some anger over losing her job.*

Better: *Manuel said he was happy in the new group, but appeared uncomfortable in the first session.*

POSITIVE AND NEGATIVE

Do not paint your client as entirely positive or entirely negative. Clients have strengths and weaknesses, and they have assets and problems. Give a balanced picture of the client, noting strengths and weaknesses, positive gains and negative problems.

AGREEMENT

There should be evidence that you and the program to which the client was referred agree on the plan for the client and that you have had interaction with each other regarding the plan. This interaction and agreement can be documented by reporting on team or staff meetings you attended at the other facility or meetings called specifically to discuss the client's goal plan or treatment.

CHANGES TO THE PLAN

Sometimes even the best plan must be changed for any number of reasons: the client gets sick, the provider is closed for snow or overcrowding, the client has a death in the family, the plan is too difficult or is not addressing the real issue. If there is a lack of progress toward the original goals:

1. Note the lack of progress in the notes.
2. Note the recommended changes to the treatment plan in the notes.
3. Revise the treatment plan.

RECORDING EXERCISES

Exercise One   Recording Your Meeting with the Client

Instructions: Your first case note will be your first meeting with the client after the treatment planning conference or disposition meeting and after developing the final plan and referral options. In this meeting, you go over the plan with the client, guardian, or the client's parents, in the case of a child, and note the client's response to your proposed plan. Also note any changes you make as a result of this interview.

Write a note for the record of the client whom you are following. This note will indicate that you met with the client and discussed the service plan. Use the four elements for writing notes. This first note is generally a little longer than the others—perhaps six to twelve sentences in all.

You can practice in the space provided here; then, when you complete the note satisfactorily, rewrite it into the chart, using the page in the Appendix marked "Contact Notes."

Exercise Two   Recording a Client Contact: Part A

Instructions: Read the following material, and then write a paragraph of no more than six sentences that gives:

1. The focus of the interview
2. Your assessment based on a concise summary of behavior, appearance, affect
3. Any resolution that takes place
4. The reason for the next contact or the follow-up that will occur

Mrs. Pell is seen in the emergency room, after which you are called by the ER physician. He tells you that her friend brought her in and that she arrived complaining of chest pains and shortness of breath. She was extremely anxious; and during her physical examination, she confided in him that she is suffering physical abuse at home and is afraid. He is uncomfortable discharging her from the ER until you have seen her. She is about 26 years old, intelligent, and a bit unkempt. You notice old bruises on her arms.

You: How are you?

Mrs. P: I guess Dr. Ingram told you—a little scared.

You: (sitting down beside her) He said that you were facing some problems at home.

Mrs. P: I am. (avoids looking at you)

You: Can you tell me a little bit about that?

Mrs. P: (nods)

You: Where would you like to start?

Mrs. P: (speaking just above a whisper) Well, I, I can't seem to get along with my husband.

You: Sounds like there are some problems.

Mrs. P: Well, we seem to fight all the time. (tears well up) I love him. I really do, but he doesn't believe me!

You: Do you feel like you could talk about some of those fights?

Mrs. P: (nods and reaches for a tissue) Yeah. I'm going to have to. I just don't know how to begin. It's gone on so long.

You: Maybe you'd like to start with what's been going on recently.

Mrs. P: Well, recently things have gotten so much worse. I feel as though it's something I'm doing. I think my husband has a short fuse. He had a head injury as a child, and he blames his temper and his moods on that.

You: So he can be pretty moody.

Mrs. P: Oh yes! And I get the brunt of it. I try to remember that and be careful—not upset him. I just feel like I'm walking on eggshells all the time lately. He just goes off at me at the least little thing.

You: (gently) Can you give me some examples?

Mrs. P: Well, two days ago I forgot to get a roast at the store. I made a meat loaf for dinner instead. He wanted roast, and he just went off when he got home. He wouldn't eat, threw the dishes on the floor, and turned over the table and then held my head under the tap in the kitchen sink. I could hardly breathe. He told me he'd let me up when I agreed to fix exactly what he asked for from now on.

You: And you agreed.

Mrs. P: (nodding) Yes. I just don't have a choice with him. I have been thinking for years—well, we've been married six years—that I could help him with this. I even thought I could become so important to him that he would never hurt me. But in all this time things have gotten worse.

You: I notice your friend brought you in. Are there people you can turn to?

Mrs. P: My friend only suspects what goes on. We never talk about it directly, but I'm pretty sure she knows. She told me one time I didn't have to stay in a bad marriage, and she looked at me kind of funny when she said it. I just think she has her suspicions.

You: So you really don't feel comfortable opening up to her about this.

Mrs. P: I don't feel comfortable opening up to anyone. My family told me not to marry him. He was a loner. He didn't like to be around them, and they thought he would keep me from seeing them. They were right, but for all these years I've pretended they were wrong.

You: You felt bad that their predictions turned out to be so true.

Mrs. P: That's right. He doesn't like them. About a year after we were married, he told me I couldn't see them again, and I've been sneaking around ever since. I make excuses to them—why we're not there at Christmas or why he didn't come along. They probably know.

You: It sounds like your world is getting tighter and tighter.

Mrs. P: Well, as I try to do everything he demands, it is. I have no friend except Suzanne (nods toward the waiting room), and I spend a lot of my time doing everything for him.

You: Do you work outside the home?

Mrs. P: Well, see, I used to, but he put a stop to that. He said the company I worked for was putting ideas in my head after the second promotion I got there. I think he was jealous, but he made me quit and said no wife of his was going to have a better job than he has!

You: Did Dr. Ingram talk to you about stress?

Mrs. P: Yeah. He said he thinks there is nothing serious wrong. But he could see these bruises, and he asked about them and then he called you. He was very nice, and he said I needed to look at the way things were going so I didn't keep having these episodes. (blurts out) I wish I didn't have to go home!

You: You don't have to. I can easily arrange for you to go into a shelter, and that would give you time to think some of this through.

Mrs. P: If I did that, I could never go home again. He's likely to kill me for seeing the people here. He doesn't like it when I see people he doesn't know. If I actually left and stayed somewhere for awhile, he'd never forgive me.

You: He means a lot to you.

Mrs. P: He's all I have. I, I, I've just started to think maybe I should leave. I'm just so afraid to. Afraid of what he'd do and afraid to be on my own. (looks pensive) He won't find out from the hospital that I saw you?

You: He shouldn't. I'll speak to them out there before I leave. Your time with me is absolutely confidential.

Mrs. P: I better get out of here. I'm afraid he'll find out I was here and—thanks. Really, this has made me feel better.

You: I think you should go too if you are concerned. Let me ask you one thing because I'm concerned about your situation at this point. How do you think I can continue to help you?

Mrs. P: (pulling her clothes on, looks over at you startled) You want to stay in touch?

You: I'd like that. My agency stays in touch with any number of women in your shoes, and we have a support group for women and a hot line if you need us quickly. Is there some way you think we could help you?

Mrs. P: (brightening) I didn't think about staying in touch with you. I could do that. I could even come in. Suzanne is bringing me here on Tuesday morning for an EKG. Could I see you then?

You: I would be happy to see you then. When is your appointment?

Mrs. P: I have to be here at 10:30. You could meet me in the waiting room out there at 10:20 and go up with me. We could talk more about what I should do. Sort it out.

You: I'll be there. Could you think a little bit about our shelter in the meantime? See if you think that is something you might want to do if things get too bad.

Mrs. P: They just might. I, well, it can't continue like this. I've told him I love him. He doesn't hear me. Sometimes I think he'll kill me someday, and I say to myself, "What are you waiting for? Run. Get out of here." But I have no place to go, and like I told you, I can't turn to my family. Oh, they'd help me in a minute, but he'd look there first, and they are no match for his anger.

(Mrs. P has her coat on and her hand on the door handle.)

You: I will be here Tuesday morning. If you need me in the meantime, or the agency, call us (handing her a business card). Tuesday we'll talk some more and see if there are some other things we might be doing to help you with this.

Mrs. P: (smiles, extends her hand) Thanks! (calling over her shoulder) I promise to call if I need anything!

Write your case note here:

Recording a Client Contact: Part B

Instructions: Read the following material, and then write a paragraph of no more than six sentences that gives:

1. The focus of the interview
2. Your assessment based on a concise summary of behavior, appearance, affect
3. Any resolution that takes place
4. The reason for the next contact or the follow-up that will occur

Mr. Dudley comes to your office in a rumpled plaid shirt. He has oily hair. He appears to have neglected his appearance, which is not typical since he stopped drinking. He has been in an outpatient alcohol treatment program for four weeks. He sits in the chair beside your desk and appears sad and ready to weep.

You: How are you?

Mr. D: Not so hot.

You: Can you tell me a little bit about what's going on?

Mr. D: I quit the treatment program.

You: Can you tell me something about what happened?

Mr. D: Well, I got tired of going. I figured I could do this myself.

You: So you wanted to try not to drink on your own.

Mr. D: Yeah. I thought most of the people in that program were whiners and complainers. All the time whining about how hard life had been, and I got tired of listening to it.

You: So you went out on your own.

Mr. D: Yes.

You: How did that work?

Mr. D: Not so hot.

You: I can see that. Could you tell me a bit more?

Mr. D: What does it look like? I started drinking again. There isn't a whole lot *more* to tell.

You: Just like that. You left the program and started to drink?

Mr. D: Well, no. I was sober for about four days. Went to work and everything. Stuff was happening at work. Some guys got laid off, and I thought I might be next.

You: So you decided to drink.

Mr. D: (hesitantly) Well, yeah.

You: I'm not clear how drinking is related to your concerns about the layoff.

Mr. D: Oh, c'mon. I was upset. I thought I might be next. I don't know. It all happened so fast. I just thought I'd stop with the fellows for one drink on the way home from work, and one thing led to another, and I've been drinking . . .

You: (without sarcasm) Looks like for several days.

Mr. D: Yeah, I can't stop.

You: What's going on with your job?

Mr. D: The first day I called in sick, but today I didn't do anything. I came here because I'm afraid.

You: Sounds like you want to stop this binge.

Mr. D: I do. I didn't know where else to go.

You: It looks like you could use a detox unit.

Mr. D: I could use getting away from bars.

You: We can talk about a plan for that when you are sober. What do you think of going into detox?

Mr. D: Yeah. Well, I suppose the missus and the boss would appreciate that.

You: What about you?

Mr. D: That's why I came here. I can't stop drinking. I'm a loser.

You: You must be feeling pretty bad about yourself.

Mr. D: Well, I screwed up again. Wouldn't you?

You: It's hard to feel like a loser. We can't address that while you are drunk, but after you sober up I will be happy to talk to you about making some changes.

Mr. D: Haven't we been through all this before?

You: We only started this five weeks ago. I don't think we have. I would like to see you go to the detox unit for seven days.

Mr. D: Yeah, I will.

You: Good. I'm not willing to talk to you about changes when you've been drinking. I'll call First Step Detox, and then I will see you there during your stay so that we can plan what happens next.

Mr. D: Thanks.

Write your case note here:

Exercise Three   Using Government Guidelines
to Correct Errors

Instructions: Using the section in this chapter on the common requirements of government funding sources, correct (or change) the following case notes.

Case Note #1: Winnie was at the hospital for tests on 3/17/99. Those tests were done to determine whether or not she has a tumor on her thyroid gland. Homemaker assistance has been arranged.
(The date for the tests is wrong. It should read 2/13/00. Correct the note accordingly in the space provided.)

Case Note #2: Carmela is attending the partial hospitalization program three days a week. Today she appeared brighter and more talkative. Our interview focused on her need to find a more independent living arrangement. Partial staff will assist her in looking at supported living programs in the community.
(Carmela is actually attending the partial program five days a week. Correct the note accordingly in the space provided.)

Exercise Four   Spotting Recording Errors

Instructions: Tell what is wrong with each of the case notes that the workers wrote in the following situations.

Case #1: Jim came into the office today looking depressed. He said he wants another job. He sat slumped over in a chair and was unkempt. The worker wrote:

1/17/97 (Office Visit): Jim came into the office today to see about getting a different job. He is not working at present. Will call Goodwill to see if they can place him temporarily at the bakery where he was before.

Case #2: Alice has been asking to be relocated to another group home since December when another client, Cheryl, moved in. Alice and Cheryl have fought ever since. The staff is not sure which client should move, and they have communicated their concerns to the case manager. The worker wrote:

3/8/97 (Phone): Alice is carrying on again about her housemates. She is trying to get a better housing assignment. Will call the house where she is staying and see if something can be done.

Case #3: Kitsu is attending an intensive outpatient rehabilitation program for his drinking. He sees his case manager at the site about once every month. Recently it was decided that Kitsu is not making the progress he was expected to make. Part of this is due to his job, which he says prevents him from coming to outpatient regularly. In order to accommodate his night schedule at work, his services will now be given in the early evening before he goes into work. The worker wrote:

8/6/97 (Site Visit): Met with Kitsu and his therapist at the rehab program. Therapist is concerned about Kitsu's lack of attendance. Changes will be made in his program to facilitate attendance.

Chapter 24

Developing Goals and Objectives at the Provider Agency

INTRODUCTION

In this chapter, you will step out of your role as case manager and into the role of the person primarily responsible for implementing the client's service plan at the provider agency. In the agency where the service is actually given, the provider agency, goals and objectives are written very specifically and in greater detail. Here the broad general goal supplied by case management is broken down into more specific goals and objectives. This enables the staff at the agency to know exactly what they are planning for the client.

When the referral arrives at the treatment or service agency, it is that agency's staff's turn to look at the case manager's stated goals and objectives for the client and decide just how to meet those in the time allotted. Completion of these more specific goals is expected to take place during the time for which the case manager has authorized payment for services to the client. Sometimes the client cannot meet the goals in that time or needs more time due to other issues that have surfaced or new problems that have occurred. For example, a client who has periodic difficulty with asthma was hospitalized on a pulmonary unit for a week and missed several weeks of services, necessitating an extension to the agreement. In another case, a client did not do well in the program where she went four days a week to learn more about independent living. Although she appeared to make progress, her progress was slower than anticipated, so the case manager extended the authorization for six more weeks. In these cases, the case manager authorized additional time for the client in that agency.

Much of the material in this chapter is based on the work of Arnold R. Goldman from his newsletter *Practical Communications*.

EXPECT POSITIVE OUTCOMES

Goals are actually the outcome you expect to occur as a result of the treatment, service, or intervention you have chosen. Goals are written, therefore, in the positive—what *will* happen, rather than what will *not* happen or what *might* happen.

To write a positive goal, state what will happen as a result of the intervention or service being provided. Use the word *will* in constructing your goals.

WRITING THE GOALS

First write the goals for the client. (Each client will have more than one goal.) In broad general terms, state what it is you intend to bring about:

> *Phyllis will be able to play cooperatively with the other children.*
> *John will be able to abstain from drinking.*
> *Gladys will have nursing care in her home.*
> *Harriet will have reliable housing.*
> *Paul will complete job-readiness training.*
> *Michael will complete a drug treatment program.*
> *Agnes will have regular medical checkups for her asthma.*

Note

- Your goal is the result of the treatment or intervention.
- The client is the subject of the goal. (The therapist, case manager, or treatment team is never the subject.)
- There is only one condition for your goal.
- The goal is written in one sentence.
- The goal is written in the positive.

The following examples demonstrate ways goals might be worded:

Poor: *Johnny's grandfather will get along better with Johnny.* Johnny's grandfather is not in the program, nor does Johnny live near his grandfather. It is unlikely, therefore, that the program will affect relationships among relatives that do not live in Johnny's home and have no connection to the program. Look instead at relationships more likely to be affected by your intervention.

Better: *Johnny will work cooperatively with other children at school.*

Poor: *To help Anne relate better to others.* This makes the therapy team the object of the goal.

Better: *Anne will work with the other workers without arguing.*

Poor: *Alice will communicate verbally and attend AA meetings.* Here there are two goals that are unrelated in the same sentence.

Better: *Alice will communicate verbally. Alice will attend AA meetings.* (two separate goals)

Poor: *Horace will attend work consistently. He will accomplish this by August 6, 2001.* Two sentences are cumbersome.

Better: *By August 6, 2001, Horace will attend work consistently.*

Poor: *Alice will no longer be alone in her home.* Here the goal tells what Alice will not do.

Better: *Alice will receive nursing care in her home for her medical problems.* Here the goal tells you what Alice will gain or learn.

Poor: *We will refer Paul to public housing and hope they can help him.* Here the goal sounds doubtful and contains an editorial note that has no place in the goal.

Better: *Paul will obtain housing for his family.*

OBJECTIVES

Goals are often quite similar for similar populations. Many clients seek help for similar reasons and, therefore, may have similar goals. ==*What individualizes a client's plan are the goal objectives,* often called "treatment objectives."==

==Every goal has objectives. The objectives are either concrete, observable, and measurable manifestations of a treatment goal or the individual steps to achieve the treatment goal.== Here are some examples of these two types of objectives for three different clients.

1. Billy's objectives:

Concrete, Observable, Measurable Manifestations

Billy will talk to at least three other children everyday without hitting.

Individual Steps

Billy will observe the activities without hitting other children.

Billy will participate in part of at least one activity daily without hitting other children.

Billy will complete one activity daily without hitting.

Billy will participate in more than one activity without hitting.

Billy will participate in an entire day of camp without hitting other children.

2. Karen's objectives:

Concrete, Observable, Measurable Manifestations

Karen will attend AA meetings.

Individual Steps

Karen will determine where AA meetings are held.

Karen will choose the site most convenient for her.

Karen will attend one AA meeting as an observer.

Karen will participate in an AA meeting.

Karen will attend two AA meetings a week.

3. Katherine's objectives:

Concrete, Observable, Measurable Manifestations

Katherine will go to the senior center twice a week.

Individual Steps

Katherine will call her local senior center for hours.

Katherine will choose the dates and times she prefers to go to the center.

Katherine will arrange for transportation with the case manager's help.

Katherine will attend the senior center twice a week.

HOW TO IDENTIFY THE CLIENT'S STRENGTHS

Clients have strengths that may be useful in working toward the goals. Be sure to use these strengths in planning for your client. When you are looking for the strengths of the client, here are some factors to check (these apply to both children and adults):

1. Look for supports in the community for the client and the client's family. For example, is the client involved in a church, a club, or a recreational program? Are there any other social services involved? Does the client have a strong circle of friends at work or strong ties to a community?

2. What are the religious or cultural beliefs, practices, or values observed by this person, and how does he or she use these for support and comfort?

3. When your client interacts with others, such as staff, family, and pets, what interpersonal skills does he or she appear to have?

4. What special abilities or skills does the client possess?

5. If you gave the client a choice as to what he/she would prefer to do, what would he/she be most likely to choose? This is particularly important with children.

6. If the client has contact with his or her family, what does the family do together? Do they eat together, go to church, meet for special occasions, or watch TV?

7. What hobbies, recreational activities, or talents does the client pursue? What interests this person?

8. What activities, people, or groups give comfort to your client? Does your client have a pet?

9. With whom is the client most likely to want to spend time? This is important in regard to children.

10. Who outside the client's family has shown an interest in this person?

COMBINING GOALS AND TREATMENT OBJECTIVES

Think of goals and objectives as being part of a continuum ranging from abstract to concrete. Figure 24.1 shows where goals and objectives would fall along such a continuum. You can apply the "See Billy" test to your goals and objectives. Read what you have written, and ask yourself if you will actually be able to see or hear the client doing that. If you cannot, you have a goal; but if you can, you have an objective. You can see or hear the client achieve the objectives. Here are some things you could not hear or see, and so you would not use these as objectives:

1. *The client will gain insight into his problems with his mother.*
2. *The client will understand the importance of AA.*
3. *The client will work well with other children at school.*

Instead, objectives in these areas would be written as follows:

1. You might know the client gained insight by what he talks about. So you might write a treatment objective that reads: *The client is able to talk about the problems he and his mother have.*

Goals are more abstract, and objectives more concrete

Abstract   Concrete

Most goals fall here   Objectives fall here

Figure 24.1

2. You could see the client's understanding of the importance of AA by the way the client attends meetings. So you might write a treatment objective that reads: *The client will attend AA twice a week.*
3. You could hear that the client worked well at school. So you might write a treatment goal that reads: *The client will receive positive reports from his teacher that he cooperated with the other children on classroom projects.*

FINISHING TOUCHES

Proper Endings

End every goal statement with one of these phrases:

"as manifested by . . ."
"as evidenced by . . ."
"as demonstrated by . . ."
"as indicated by . . ."

Then write the objective after that phrase. For example:

Goal A: Anita will understand the importance of AA *as evidenced by* . . .
Objective 1: attendance at AA meetings twice a week for six weeks.

Numbering System

Develop a numbering system for easy identification so you can tell which objectives go with which goals.

Make goals A, B, C, D, and so on.
Make objectives 1, 2, 3, 4, and so on.

Thus, you will have an A1 that represents goal A, objective 1. In this way, you do not have to write out everything in your notes; you have an easy reference system with which to refer to goals and objectives. For instance, in the example of Anita, you might want to note progress in Anita's consistent attendance at AA meetings. You would say that Anita is meeting A1 by having attended AA meetings twice weekly for two weeks.

Every goal must have at least one objective. It can have more than one objective, as you see fit.

Target Dates

Give each objective a target date. Remember to set the date for the amount of time you are willing to try this particular intervention without getting the desired result. In this way, as the worker, you will be able to monitor if someone is progressing. With no target date, it is easy to leave people in programs and treatment for unnecessarily long periods of time. Even if the client has not reached the goal by the target date, you can assess any progress that has been made and determine whether the program is working and needs a new target date or whether the program is not working and other interventions need to be tried or other arrangements need to be made for the client.

If the client meets the goal or the objective before the target date, simply record this in the progress notes. Also note the new goals and objectives and the new target dates for these as the client continues along an improving continuum.

In some cases where clients have a chronic mental illness, some form of mental retardation, or problems related to aging, a provider of service may be asked to provide an intervention to simply prevent the client's condition from growing worse. In this case, you set the target date according to the date at which you would expect to see deterioration if your intervention was not working. When you reach the target date, if the patient's behavior or situation has remained stable, you can extend your target date. Review the intervention and the person's stability at each target date before extending it, and note your review and extension in the chart.

Treatment Interventions

If you give the client an assignment or some task to do that will help him or her complete one of the objectives, this is a treatment intervention. It is not a goal or objective. It is a treatment intervention because it is what you (the worker) are doing to help the client complete an objective and work toward his or her ultimate goal. In the earlier example, you might ask Anita to read some testimonials of people who found AA meetings particularly helpful.

Notice that a *treatment objective* gives some anticipated change in the client's behavior (for example: Client will attend AA meetings twice a week), while an *intervention* tells what you, the case manager, will use to bring about this change (for example: Worker will give client testimonials to read).

Long Term and Short Term

Sometimes you are asked to state what your targets are for the long term and for the short term. This may happen if you are in a meeting regarding the client or receive a letter from the client's insurance company. Simply state your goals as your long-term targets and your objectives as your short-term targets.

The goals and objectives shown in Figure 24.2 are for a little girl at a therapeutic camp program. When case managers make a referral to another program, the program generally sets the objectives. Goals are developed in broad general terms at the case management unit, and the service provider develops the more specific goals and objectives. Or the staff of a service provider can take the broad general goals from case management and develop objectives for them. For example, a person referred to a partial hospitalization program would be referred there because of the need to attain a particular goal, such as learning to control anger. The objectives would be designed by the staff there. A person referred to AA would work out the objectives with his case manager because there is no paid staff at AA. Exactly who designs goals and objectives will depend on the program chosen for the client and the extent to which that program has in-house case management services.

CLIENT PARTICIPATION/COLLABORATION

We do not make goals for clients without collaboration with the client. If you are working with a child, you want to note that the parent or parents were involved in the decisions. For an adult, you need to include the fact that the adult has participated in determining his/her own goals.

On the assessment and evaluation forms, there are usually places where clients are asked what they see as the main issues to be resolved and what they expect of services. Each of these forms addresses this in a different way, but look at this material when developing the goals. The information you collect from the client should indicate that you and the client looked at this matter together.

EXERCISES

Instructions: Use the following vignettes to develop detailed treatment plans for the clients.

Helen is 89 years old. She has lived independently in her own home until now. Last week she fell off a chair in her home while reaching up to clean out the top shelves in one of her closets. She sat with her arm aching all night until morning when she called her daughter to take her to the hospital. There an x-ray was done, and the arm was set. The

---

Example of a hypothetical camper's treatment goals and objectives

Goal A   Sherry will play cooperatively with the other children as evidenced by

1. Sherry will be able to engage in one group activity daily without yelling and pushing the other campers by the end of camp.

2. Sherry will be able to follow her team leader in games each day by the end of camp.

Goal B   Sherry will demonstrate her understanding of leadership as evidenced by

1. Sherry will lead one activity with her team every week from the third week of camp to the end.

2. Sherry will talk in group about what it takes to be a good leader.

3. Sherry will assist younger children in a team activity at least once during the camp session.

Goal C   Sherry will express anger verbally as evidenced by

1. Sherry will tell another about her anger at least twice during the camp session.

Figure 24.2

hospital staff sent her home; and her daughter, who works, called the Office of Aging to help her plan so that her mother could remain at home. The Office of Aging set, as the main goal, "To continue living in her own home with assistance" and referred the case to your agency to provide the service. Helen is unable to make meals or bathe herself, and the pain medication is making her dizzy. The Office of Aging has called your agency, which provides homemaker and in-home care, to provide the actual service to Helen.

1. The broad goal is *to keep Helen in her own home*. List two more specific goals your agency will work on with Helen. Next, place the objectives for each goal under that goal. Remember the phrase "as evidenced by . . ."

    Goal A: _____

    _____

    Objective 1: _____

    _____

    Objective 2: _____

    _____

    Goal B: _____

    _____

    Objective 1: _____

    _____

    Objective 2: _____

    _____

2. Identify Helen's strengths, and show how these strengths will help you meet the goals.

    Helen's strengths:

    _____

    _____

    _____

    _____

    Explain how these will help her meet the goals:

    _____

    _____

    _____

    _____

    _____

    _____

    _____

    _____

3. Helen broke her arm on November 9th. What are your target dates for your goals?

_____

_____

_____

_____

_____

_____

_____

4. Describe an intervention you will use with Helen to help her meet her goals.

_____

_____

_____

_____

_____

_____

_____

_____

Art has been chronically ill with schizophrenia for many years. It started when he was in college, and he has been unable to hold a job. His family has just moved to your area. Art's sister, with whom he lives, has gone into the mental health center seeking services for her brother. Art's parents are deceased. The sister states the move has upset Art, and he appears ready to have another acute episode. She is also concerned that he is not taking his medication as he should, further jeopardizing his health. The sister hoped the case manager could find a place to send Art for treatment following the move. In college Art majored in engineering, and he is quite good at math. Art has been referred to the partial hospitalization program. There you are responsible for devising goals and objectives for Art.

1. The broad goal is *to help Art adjust to the move.* List two more specific goals your agency will work on with Art. Next place the objectives for each goal under that goal. Remember the phrase "as evidenced by . . . "

Goal A: _____

_____

Objective 1: _____

_____

Objective 2: _____

_____

Goal B: _____

_____

Objective 1: _____

_____

Objective 2: _____

_____

2. Identify Art's strengths, and show how these strengths will help you meet the goals.
   Art's strengths: _____

_____

_____

_____

_____

   Explain how these will help him meet the goals:

_____

_____

_____

_____

_____

_____

_____

3. Art and his sister went to mental health case management on January 12th. What are your target dates for your goals?

_____

_____

_____

_____

_____

_____

4. Describe an intervention you will use with Art to help him meet his goals.

_____

_____

_____

_____

_____

_____

✗ Chapter 25

Terminating the Case

INTRODUCTION

We have looked at how clients enter the human service system and how their services and treatment are determined and monitored. There is usually a point where clients leave the system, moving on in their lives. Here are some of the reasons a case is terminated.

1. *The client and the case manager agree the client is ready to move on.* This is the ideal. The service or treatment has been successful and is no longer needed. Many clients do leave for this reason, feeling that their original issues and problems are less significant than they once were or that these problems have been resolved.

2. *The client dies or moves away.* When clients die or move to another jurisdiction, their cases are closed. If they formally request that their records be sent to the new jurisdiction, this should be done immediately in order to facilitate a smooth transition to the new program.

3. *The funding source will no longer finance services.* Managed care has introduced limitations on care that you and the client may find unrealistic. It is important to work with the client to find alternatives to your service. Support groups or specialized programs funded by other sources may not give the level of service you have provided, but may help the client make the adjustment.

4. *The client no longer wants the services.* Clients may be dissatisfied with the services being offered and request that their cases be terminated. In situations like this, sit down with the dissatisfied clients and learn why they are not pleased with the service. This may provide you with valuable information about how you or a provider agency are perceived by clients, and it may facilitate clients leaving with the feeling that they can come back if they need to do so.

Not all clients will leave case management. A child with autism may need services all his life. A woman with severe developmental disabilities may require case management during her entire lifetime in order for her to live in her community successfully.

A SUCCESSFUL TERMINATION

Cases should not be closed without the client, or the client's family, if they are involved, knowing that this is about to occur and why. A letter by itself cannot convey warmth and concern for the client and often comes across as bureaucratic and unfeeling. A phone call is not much better. You might convey warmth, but there is no meaningful exchange or documentation.

All clients, except those who have moved away or died, whose cases are being closed should receive two things from the agency:

1. An opportunity to meet with you to discuss the termination
2. Followed by a letter that outlines the main points in your interview and invites the client to return if the need arises

The Final Interview

Leaving anything can be difficult. This is particularly true for clients who may have grown fond of the people at the agency or who may feel uncertain about how they will handle life on their own. Sometimes these terminations are milestones. The clients have reached a new level of independence, emotional health, or sobriety. But the accomplishment is tinged with misgivings. Try to recognize the underlying feelings your client is experiencing in the final interview and respond empathically.

Use the interview to summarize the reasons for the termination. If there are accomplishments, go over how far the client has come since first seeking help some time before. If clients have requested their cases be terminated themselves, ask them to tell you more about their reasons for making this request and be open to their suggestions for change. Invite questions. Clients often want to know where they should turn should old problems resurface. Give information the client can use, particularly information that will help the client prevent a relapse or regression.

There should be a sense of reassurance during the interview that clients are not being cut off or dumped and that they are welcome to return should they need services again. You may not need to say all of that explicitly. Nevertheless, clients may feel dumped or dropped and be unable to express that to you. Speak to these concerns in the final interview.

The Letter

The letter is a follow-up to the final interview. It should summarize the main points of the interview, recapping briefly what was discussed. There should be a brief statement as to why the termination took place so the client has documentation of this. In addition, questions that seemed particularly important to the client during the interview should be answered again in the letter. This is especially important if the client needs addresses, names of resources, and other supportive information.

Documentation

Just as with all contacts with the client, this last contact and letter should be documented in the client's chart. In the note, the focus of the interview would be the termination of the case, and you would go on to note the highlights of the discussion, any follow-up arrangements that might have been made for the client, and the client's response to the interview.

THE DISCHARGE SUMMARY

In most cases, your agency will ask for a termination summary or discharge summary. This is not the same thing as your final case note discussed earlier. Although you would include the information from that contact note, in the discharge summary, you are actually summarizing the most important information about what took place while the client was working with you. It is wise to think of your summary as a document that may go to other professionals who will see the client in the future. What you include in your summary should be helpful to a new person developing a strategy to help the client. For this reason, your summary should discuss what was tried, what worked, and what was less successful and why.

In addition, your discharge summary is also a public relations tool for your agency. This is a major way that other programs and professionals can know the quality of the care you give in your case management unit. The summary provides a view of the excellence of your services. Sloppy summaries with little useful information or written to indicate little organized effort on behalf of the client can make your agency look unprofessional. Your agency may have a standard format for discharge summaries that you can use as a guide to writing good discharge summaries. If not, here are important items to include:

1. The major presenting problem that brought the client to you
2. Your goals and objectives for the client
3. The extent to which the client participated in formulating these goals and objectives
4. Progress that was made or goals that were accomplished
5. Problems that were identified but were not addressed
6. Diagnoses
7. Any medication that might have been prescribed by physicians working with the case management unit (Be sure to note the medication by name, the dosage, the frequency, and any adverse reactions.)
8. How the client appeared to be at intake and how the client appeared to be at termination

CONCLUSION

Termination should be approached as skillfully as you approach all other aspects of case management. When cases are closed well, clients and the agency both benefit. Clients feel reassured and supported as they take leave of your services. In the community, the perception of your agency as a caring and professional place is strengthened.

TERMINATION EXERCISES

Exercise One

Termination of a Middle-aged Adult

Instructions: 1. Record in your contact notes a termination interview, indicating the reason for termination and any follow-up arrangements you made for the client after he or she leaves your services. Refer to the goals for such an interview in this chapter.

2. Write your client a letter as a follow-up to this interview summarizing for the client the discussion and resolutions that took place in your last contact.

3. Prepare a discharge summary on your client that will be the last item in this client's chart. Use the guidelines for a discharge summary found in this chapter.

Exercise Two

Termination of a Child

Instructions: 1. Record in your contact notes a termination interview, indicating the reason for termination and any follow-up arrangements you made for the client after he or she leaves your services. Be sure your termination interview includes the parents and indicate if the child was present. Refer to the goals for such an interview in this chapter.

2. Write your child's parents a letter as a follow-up to this interview summarizing for them the discussion and resolutions that took place in your last contact.

3. Prepare a discharge summary on this child that will be the last item in this client's chart. Use the guidelines for a discharge summary found in this chapter.

Exercise Three

Termination of a Frail, Older Person

Instructions: 1. Record in your contact notes a termination interview, indicating the reason for termination and any follow-up arrangements you made for the client after he or she leaves your services. Indicate if others were present, such as relatives or a worker from a new service where your client will be going. Document where you conducted the interview. Refer to the goals for such an interview in this chapter.

2. Write your client a letter as a follow-up to this interview summarizing for the client the discussion and resolutions that took place in your last contact. If you have the client's permission to do so, indicate that you are sending a copy of your letter to the person who was present for the termination interview.

3. Prepare a discharge summary on your client that will be the last item in this client's chart. Use the guidelines for a discharge summary found in this chapter.

Wildwood Case Management Unit Forms

References and Reading List

American Psychiatric Association. (1994). *Diagnostic and statistical manual of mental disorders* (4th ed.). Washington, DC: Author.

Bednar, R. L., Bednar, S. C., Lambert, M. J., & Waite, D. R. (1991). As quoted in G. Corey, M. S. Corey, & P. Callanan, *Issues and ethics in the helping professions* (1993). Pacific Grove, CA: Brooks.

Beisser, A. (1970). As published in Fagan & Shepherd, *Gestalt Therapy Now.* Harper Colophon.

Burns, D. D. (1980). *Feeling good: The new mood therapy.* New York: William Morrow.

Caudill, O. Brandt. (1996, October). *The California Psychologist,* p. 4.

Corey, G., Corey, M. S., Callanan, P. (1993). *Issues and ethics in the helping professions.* Pacific Grove, CA: Brooks/Cole.

Dinkmoyer, D., & Losoncy, L. E. (1980). *The encouragement book: Becoming a positive person.* Englewood Cliffs, NJ: Prentice-Hall.

Goldman, A. R. (1990). Special focus on basic rules of writing treatment goals and objectives. *Accreditation and certification, Vol. IV,* (3), 1–9.

Gordon, T. (1970). *Parent effectiveness training: The tested way to raise responsible children.* New York: David Mackay.

Gudykunst, W. B., & Kim, Y. Y. (1997). *Communicating with stranger: An approach to intercultural communication.* New York: McGraw-Hill.

Jackson, S. W. (1992). The listening healer in the history of psychological healing. *American Journal of Psychiatry, 149*(12), 1623–1632.

LaBruzza, A. L. (1994). *Using DSM IV: A clinician's guide to psychiatric diagnosis.* Northvale, NJ: Jason Aronson.

Lukas, S. (1993). *Where to start and what to ask: An assessment handbook.* New York: Norton.

National Organization for Standards in Human Service Education. (1967). *Ethical standards of human service professionals.*

Siegel, M. (1979). As quoted in G. Corey, M. S. Corey, & P. Callanan, *Issues and ethics in the helping professions* (1993). Pacific Grove, CA: Brooks/Cole.

Stadler, H. A. (1990). Confidentiality. In B. Herlihy & L. B. Golden (Eds.), *AACD ethical standards casebook* (4th ed.), p. 102. Alexandria, VA: American Association for Counseling and Development.

Stadler, H., & Paul, R. D. (1986). As quoted in G. Corey, M. S. Corey & P. Callanan, *Issues and ethics in the helping professions* (1993). Pacific Grove, CA: Brooks/Cole.

Stephan, W. (1985). Intragroup relations. In G. Lindzey & E. Aronson (Eds.), *Handbook of social psychology* (3d ed., Vol. 2). New York: Random House.

Three Rivers Center for Independent Living. *Language references.* Pittsburgh, PA.

Weiner, I. B. (1975). *Principles of psychotherapy.* New York: John Wiley.

Index

C

W

TO THE OWNER OF THIS BOOK:

We hope that you have found *Fundamentals of Case Management Practice: Exercises and Readings* useful. So that this book can be improved in a future edition, would you take the time to complete this sheet and return it? Thank you.

School and address: _____

Department: _____

Instructor's name: _____

1. What I like most about this book is: _____

_____

_____

2. What I like least about this book is: _____

_____

_____

3. My general reaction to this book is: _____

_____

_____

4. The name of the course in which I used this book is: _____

_____

5. Were all of the chapters of the book assigned for you to read? _____

   If not, which ones weren't? _____

_____

6. In the space below, or on a separate sheet of paper, please write specific suggestions for improving this book and anything else you'd care to share about your experience in using the book.

_____

_____

_____

OPTIONAL:

Your name: _____ Date: _____

May we quote you, either in promotion for *Fundamentals of Case Management Practice: Exercises and Readings,* or in future publishing ventures?

Yes: _____ No: _____

Sincerely yours,

Nancy Summers